SPRINGER
LABORATORY

Heinz-Helmut Perkampus

UV-VIS Spectroscopy and Its Applications

Translated by H. Charlotte Grinter
and Dr. T. L. Threlfall

With 78 Figures and 21 Tables

Springer-Verlag
Berlin Heidelberg New York London Paris
Tokyo Hong Kong Barcelona Budapest

Professor em. Dr. HEINZ-HELMUT PERKAMPUS

Heinrich-Heine-Universität
Physikalische Chemie
und Elektrochemie I
Universitätsstraße 26.43.02
W-4000 Düsseldorf 1

ISBN-13: 978-3-642-77479-9 e-ISBN-13: 978-3-642-77477-5
DOI: 10.1007/978-3-642-77477-5

Library of Congress Cataloging-in-Publication Data. Perkampus, Heinz-Helmut. [UV-VIS-Spektroskopie und ihre Anwendungen. English] UV-VIS spectroscopy and its applications/ Heinz-Helmut Perkampus; translated by H. Charlotte Grinter and T. L. Threlfall. p. cm. Includes bibliographical references and indexes. ISBN-13: 978-3-642-77479-9 (alk. paper): DM 168.00.– ISBN-13: 978-3-642-77479-9 (alk. paper: U.S.) 1. Ultraviolet spectroscopy. I. Title. QD96.U4P4713 1992 543'.08585–dc20 92-20077

Softcover reprint of the hardcover 1st edition 1992

Typesetting: K+V Fotosatz GmbH, Beerfelden
52/3145-5 4 3 2 1 0 – Printed on acid-free paper

Preface

UV-VIS spectroscopy is one of the oldest methods in molecular spectroscopy. The definitive formulation of the Bouguer-Lambert-Beer law in 1852 created the basis for the quantitative evaluation of absorption measurements at an early date. This led firstly to colorimetry, then to photometry and finally to spectrophotometry. This evolution ran parallel with the development of detectors for measuring light intensities, i.e. from the human eye via the photoelement and photocell, to the photomultiplier and from the photographic plate to the present silicon-diode detector both of which allow simultaneous measurement of the complete spectrum.

With the development of quantum chemistry, increasing attention was paid to the correlation between light absorption and the structure of matter with the result that in recent decades a number of excellent discussions of the theory of electronic spectroscopy (UV-VIS and luminescence spectroscopy) have been published. Consequently, this extremely interesting aspect of molecular spectroscopy has dominated the teaching of the subject both in my own lectures and those of others. However, it is often overlooked that, in addition to the theory, applications of spectroscopic methods are of particular interest to scientists. For this reason, a lecture series about electronic spectroscopy given in the Institute for Physical Chemistry at the Heinrich-Heine-University in Düsseldorf was supplemented by one about "UV-VIS spectroscopy and its applications". This formed the basis of the present book.

UV-VIS spectroscopy owes its importance not least to its varied applications in chemistry, physics and biochemistry. This book aims to show how UV-VIS spectroscopy can be applied to analytical problems, to the investigation of chemical equilibria and to the kinetics of chemical reactions, including photokinetics. The theoretical section has been kept to a minimum since, as mentioned above, excellent discussions of such matters are available in the literature. The details of the equipment are also described very briefly because G. Kortüm gave an outstanding discussion of this subject in volume II of the series "Anleitung für die chemische Laboratoriumspraxis"; and its basic details still apply today.

In addition to the applications, a number of UV-VIS spectroscopic techniques are discussed. However, in this case the selection has been influenced by the author's own interests. In order to obtain experimental examples, numerous measurements have been made which might also be set as practical work for students of advanced physical chemistry.

I would like to thank my colleagues for making these measurements, and drawing the diagrams.

The English translation of the second edition of this volume is due to the stimulating interest of the "Ultraviolet Spectrometry Group", London, to whom I am very grateful.

Mrs. Charlotte Grinter undertook the translation with great interest and engagement, professionally supported by Dr. T.L. Threlfall and Dr. Grinter. I would like to express my sincere thanks to Mrs. Grinter and the colleagues mentioned above for their hard labors.

A few additions have been made to the first edition, i.e. the brief section "Chemometrics" was added by the English colleagues, some figures have been changed, others are new and the cited literature has been updated where necessary. Here also the English colleagues proved to be very helpful for which I would like to express my thanks.

Thanks are also due to Dr. Enders of the Springer-Verlag for his interest and support of the publication of the English edition.

Düsseldorf, June 1992 HEINZ-HELMUT PERKAMPUS

Contents

1 Introduction

Optical spectroscopy is based on the Bohr-Einstein frequency relationship

$$\Delta E = E_2 - E_1 = h\nu \; . \tag{1}$$

This relationship links the discrete atomic or molecular energy states E_i with the frequency ν of the electromagnetic radiation. The proportionality constant h is Planck's constant (6.626×10^{-34} J s or 6.626×10^{-27} erg s). In spectroscopy, it is appropriate to use the wavenumber $\tilde{\nu}$ instead of frequency ν. Equation (1) then takes the form:

$$\Delta E = E_2 - E_1 = hc\tilde{\nu}$$

where

$$\nu = c/\lambda = c\tilde{\nu} \; . \tag{2}$$

Absorbed or emitted radiation of frequency ν or wavenumber $\tilde{\nu}$ can thus be assigned to specific energy differences or, applying the definition of the 'term value' (energy level), to specific energy-level differences:

$$\tilde{\nu} = \Delta E/hc = E_2/hc - E_1/hc = T_2 - T_1 \tag{3}$$

$T_i = E_i/hc$ is the term value. From this definition it follows that it has dimension m^{-1} in the SI-system. However, it is still commonly given as cm^{-1}; thus, wavenumber $\tilde{\nu}$ as a term difference may be given in m^{-1} or cm^{-1}. Since the wavenumber is always given in cm^{-1} in the literature, it will also be used in this book ($1\,cm^{-1} \triangleq 100\,m^{-1}$).

For absorption spectroscopy in the ultraviolet (UV) and visible (VIS) region, this range can be characterized by the information in Fig. 1.

Within the overall range of electromagnetic radiation which is of interest to chemists, UV- and VIS-absorption spectroscopy occupies only a very narrow frequency or wavenumber region. Nevertheless, this range is of extreme importance, since the energy differences correspond to those of the *electronic states* of atoms and molecules; hence the concept of "electronic spectroscopy". Furthermore, in the visible spectral region the interactions between matter and electromagnetic radiation manifest themselves as *color*. This led the early investigators to methods of measurement, the basic principles of which still apply today.

The limits given in Fig. 1 are not fixed limits because molecules exhibit absorption below 200 nm ie. above 50000 cm^{-1}. However, this spectral region is not accessible to routine measuring techniques. The short-wave-

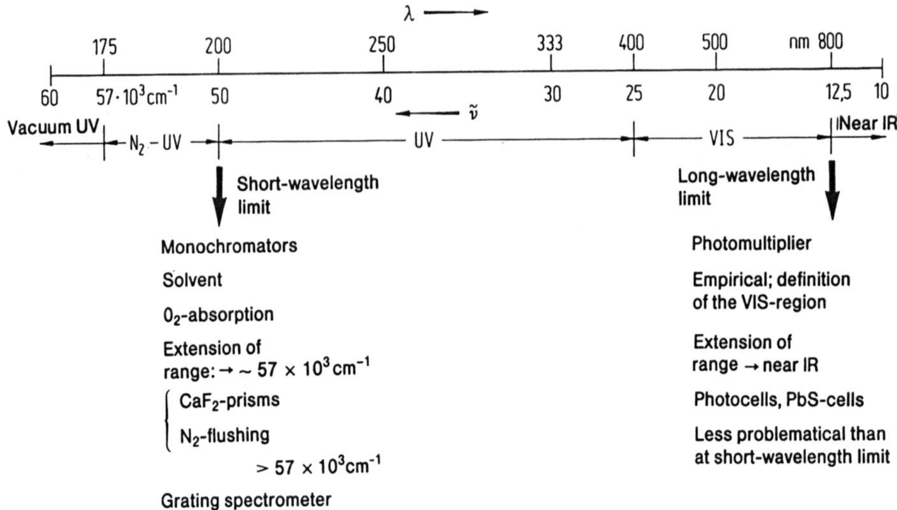

Fig. 1. The ranges of electronic spectra and their limits

length limit is restricted by apparatus and by experimental techniques. The long-wavelength limit depends less on considerations of apparatus because, apart from a few exceptions, most compounds exhibit no absorption traceable to electronic excitation in this region. There are exceptions; for example polymethine dyes, used as photographic sensitizers, and some inorganic complexes which have absorption bands that can be observed up to $2\ \mu m \triangleq 5 \times 10^5\ m^{-1}$ ($5000\ cm^{-1}$).

2 Principles

2.1 The Bouguer-Lambert-Beer Law and Its Practical Application

The Bouguer-Lambert-Beer law forms the mathematical-physical basis of light-absorption measurements on gases and solutions in the UV-VIS and IR-region [1]:

$$\lg\left(\frac{I_0}{I}\right)_{\tilde{\nu}} = \lg\left(\frac{100}{T(\%)}\right)_{\tilde{\nu}} \equiv A_{\tilde{\nu}} = \varepsilon_{\tilde{\nu}} \cdot c \cdot d \ , \qquad (4)$$

where

$A_{\tilde{\nu}} = \lg\left(\frac{I_0}{I}\right)_{\tilde{\nu}}$ is the *absorbance,*

$T_{\tilde{\nu}} = \frac{I}{I_0} \cdot 100$ in % is the *transmittance,*

$\varepsilon_{\tilde{\nu}}$ is the *molar decadic extinction coefficient.*

I_0 is the intensity of the monochromatic light entering the sample and I is the intensity of this light emerging from the sample; c is the concentration of the light-absorbing substance and d is the pathlength of the sample in cm.

Equation (4) then gives:

$$\varepsilon_{\tilde{\nu}} = \frac{A_{\tilde{\nu}}}{c \cdot d}$$

with dimensions for $\varepsilon_{\tilde{\nu}}$ of:

$1\,mol^{-1}\,cm^{-1}$ for "c" in $mol\ l^{-1}$

or

$1000\,cm^2\,mol^{-1}$ for "c" in $mol\ 10^{-3}\,cm^{-3}$.

The molar decadic extinction coefficient, $\varepsilon_{\tilde{\nu}}$, is a quantity characteristic of the substance which also depends on wavenumber $\tilde{\nu}$ (cm^{-1}) or on wavelength λ (nm).

The functional correlation between $\varepsilon_{\tilde{\nu}}$ and wavenumber $\tilde{\nu}$ is called the *"absorption spectrum"* of a compound. Since the extinction coefficient can vary by several orders of magnitude within the absorption spectrum of a single inorganic or organic compound, the logarithmic value $\lg \varepsilon = f(\tilde{\nu})$ can be used instead of $\varepsilon = f(\tilde{\nu})$ to plot an absorption spectrum [2].

The Bouguer-Lambert-Beer law is a limiting law for dilute solutions, i.e. the assertion that the extinction coefficient ε is independent of the concentration of a substance at the given wavenumber $\tilde{\nu}$ (wavelength λ) applies only to dilute solutions. ε is no longer constant for concentrated solutions but depends on the refractive index of the solution [1]. At concentrations up to $c \leqslant 10^{-2}$ mol l^{-1}, the effect is slight and lies 1 or 2 powers of ten below the usual photometric accuracy, as Kortüm showed with precise measurements on aqueous solutions of $K_3[Fe(CN)_6]$ [3].

According to Eq. (4), the application of the Bouguer-Lambert-Beer law presupposes a measurement of the relationship between the light intensities I and I_0. However, when measuring in quartz cuvettes (UV-VIS region) or cuvettes made of special optical glass (VIS region), part of the light is lost through reflection at the cuvette surfaces. In order to eliminate this source of error, a reference measurement is made in a cuvette with the same pathlength but not containing the substance to be measured. Since most UV-VIS spectroscopy is carried out with solutions, the standard cuvette contains the pure solvent, which ideally should not absorb in the spectral region under consideration.

Thus, I_0 is measured after the light has traversed the standard or reference cuvette and I after the light has traversed the cuvette containing the sample. Depending on the construction and mode of operation of the equipment, the relationship I/I_0 is shown as a value of $T_{\tilde{\nu}}$ (%) or $A_{\tilde{\nu}}$ in either analog or digital form. This result is independent of losses due to reflection and the influence of the solvent.

This still presupposes that the two cuvettes used for the measurement have the same pathlength and have been matched prior to making the measurements. Most manufacturers keep the accuracy of the pathlengths of a matched cuvette set within a few ppm. However, the continued matching of a previously used pair of cuvettes depends entirely on the care taken by the individual user of an UV-VIS spectrophotometer. In many applications, standard cuvettes are suitable; they are available in pathlengths of 1, 2, 5, 10, 20, 50 and 100 mm and, depending on the spectral region, are produced either from Suprasil or Spectrosil quartz glass or special optical glass. Furthermore, there is a wide range of cuvettes for special methods of measurement [4].

The choice of solvent depends on an adequate solubility of the substance to be measured. For example, *n*-heptane, water and trifluoroethanol or hexafluoroisopropanol may be considered as good spectroscopic solvents because they are transparent from ca. 180 nm in the UV-VIS region. However, below 200 nm the pathlength must be reduced to 1 mm and in this region the spectrophotometer must be flushed with pure nitrogen in order to reduce absorption by atmospheric oxygen to a minimum. Transmission curves for the most important solvents are shown in [2], volume 5. The UV transmission of solvents depends critically upon the solvent purity. For that reason, some manufacturers supply solvents specially purified for UV spectroscopy [5, 6].

2.2 Primary Photophysical Processes

On the basis of Eq. (3), the energy states of a molecule are summarized in an energy-level diagram.

A general energy diagram of electronic states used to explain the primary photophysical processes is shown in Fig. 2, without taking the vibrational states into account.

The individual levels correspond to the different energies of electrons in singlet and triplet states. Of all the transitions shown, standard absorption spectroscopy involves only transition I. Fluorescence and phosphorescence arise from transitions V and VI respectively.

Transitions XI, XII and XIII are non-radiative transitions which are known as *internal conversions*. Transitions XIV and XVII are *intersystem crossings*. Transition II ($T_1 \leftarrow S_0$) represents singlet-triplet absorption and being an *intercombination transition* is spin-forbidden. Thus, it occurs with a very low intensity and special methods of measurement are required to observe this transition [7]. Transitions III and IV are two-photon transitions

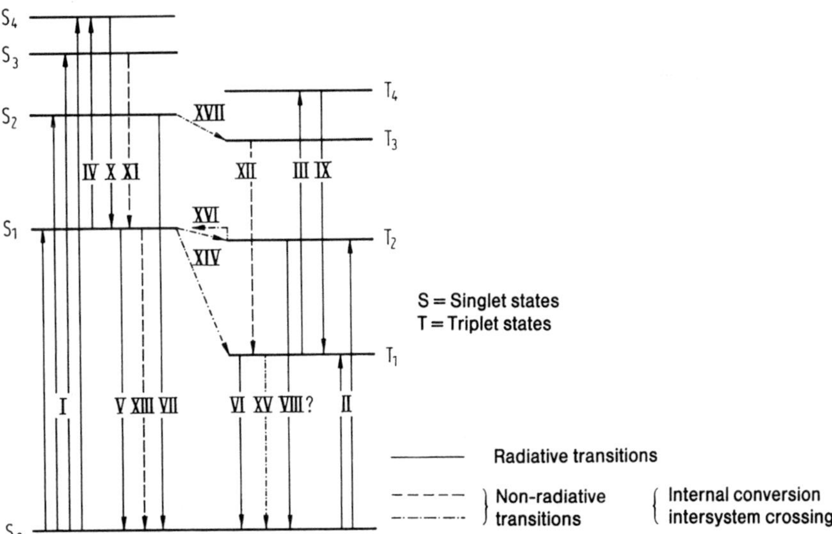

Fig. 2. General energy-level diagram for electronic excitation

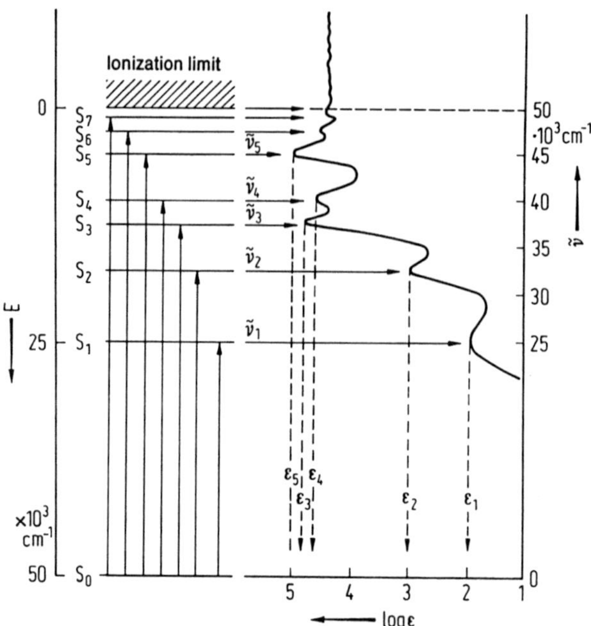

Fig. 3. Singlet-singlet transitions and their assignment to the absorption spectrum

where T_1 or S_1 must be excited primarily by the first photon. The process of resonance fluorescence is represented by VII and in practice it can only be observed in gases under reduced pressure. Birks has given an account of the primary photophysical processes [8].

Singlet-singlet transitions have been assigned to the measured absorption spectrum in Fig. 3 to illustrate that absorption maxima correspond to quite specific energy states, i.e. excitation energies. Furthermore, this figure demonstrates the important fact that, in addition to the position of the absorption maximum, the extinction coefficient, ε, is also very significant when interpreting spectra.

2.3 Vibrational Structure of Electronic Spectra

The energy level diagrams in Figs. 2 and 3 do not take into consideration the fact that *vibrational* and *rotational states* are superimposed on the electronic states. In the case of molecules having the dimensions with which we are concerned here, rotational states can no longer be resolved because the surrounding solvent molecules strongly hinder rotation in solution.

Consequently, the observed structure is caused by a superposition of vibrational states only. Figure 4 shows an energy-level diagram which for

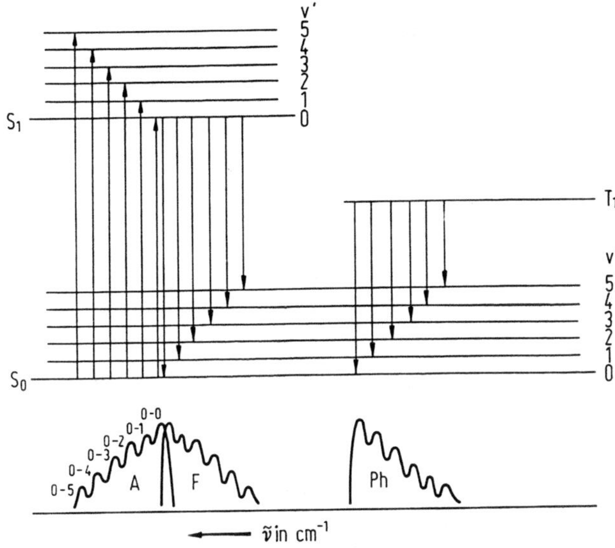

Fig. 4. Energy-level diagram including the superposition of one vibrational progression

simplification illustrates only the superposition of one vibration, i.e. vibrational progression, in each of the ground and excited states. By reference to the absorption, fluorescence and phosphorescence spectra shown in this energy-level diagram, the characteristics of such spectra can be illustrated:

a) The vibrational quanta of the excited state can be observed in the *absorption spectrum* and, in contrast, those of the ground state in the fluorescence and phosphorescence spectra.
b) Frequently, the *fluorescence spectrum* is approximately a mirror image of the absorption spectrum (for examples see [9]).
c) On account of the low energy of the triplet state, T_1, the *phosphorescence spectrum* is displaced strongly toward the red so that the fluorescence and phosphorescence spectra are normally clearly separated (for examples see [9]).

The spectra shown schematically in Fig. 4 illustrate the case of molecules which have the same geometry in the ground and excited states. This occurs only rarely, and the maxima in absorption and emission are usually displaced toward higher vibrational transitions, i.e. the 0-0 transition is no longer the most intense. The Franck-Condon principle explains this behaviour [10, 11, 12].

2.4 Electronic Spectra and Molecular Structure

The discrete molecular states assigned theoretically to the electronic states are shown in the energy-level diagram. The electronic states depend very critically upon the number of electrons in a molecule as well as on the structure or geometry and the symmetry of that molecule. Consequently, electronic spectra are an extremely valuable aid to structure determination.

The molecular eigenfunctions of the ground state and the different excited states also determine the selection rules and thus the intensities of electronic transitions. The correlation between theory and experiment may be expressed by the *oscillator strength* "f" which may be calculated theoretically and can also be established experimentally from $\varepsilon = f(\tilde{\nu})$ using Eq. (5) (see particularly Chapter 8).

$$f_{exp} = \frac{2303 \cdot m \cdot c^2}{\pi \cdot e \cdot N_L n} \int_{Band} \varepsilon(\tilde{\nu}) d\tilde{\nu} \ . \tag{5}$$

Here m = mass of an electron; c = velocity of light; e = electronic charge; N_L = Loschmidt number; n = refractive index of the solvent.

The integral represents the *"integrated intensity"* which can be simply determined experimentally. It can also be approximated by the expression

$$\int_{Band} \varepsilon(\tilde{\nu}) d\tilde{\nu} \approx \varepsilon_{max} \Delta\tilde{\nu}_{1/2} \tag{6}$$

$\Delta\tilde{\nu}_{1/2}$ is the width of the band at half its maximum intensity (fwhh).

Equation (7) gives the theoretical expression for $f_{l,k}$:

$$f_{l,k} = \frac{8\pi^2 m c \tilde{\nu}_{l,k}}{3 \cdot h e^2} G \, |M_{l,k}|^2 \ . \tag{7}$$

$\tilde{\nu}_{l,k}$ is the wavenumber of the 0-0 transition (l→k),
G is the statistical weight which equals 1 for a pure electronic transition,
$M_{l,k}$ is the transition dipole moment which can be calculated theoretically.

The transition dipole moment determines the intensity of a transition. This moment is a vector and is composed of three components in the Cartesian coordinate system. Consequently, for many planar molecules, the component vertical to the molecular plane is missing and an anisotropy of electronic excitation is present which is of great interest in molecular theory. In summary, it can be said that electronic spectra supply the following information:

1. Absorption maxima $\tilde{\nu}_{max}$ which correspond to the discrete molecular states which are strongly dependent on the molecular structure, geometry and symmetry.
2. Extinction coefficients ε_{max}, or the integral absorption over an absorption band, which give the magnitude of the transition dipole moment and are also dependent on geometry and symmetry.
3. The structure within an absorption band or within the fluorescence or phosphorescence spectrum supplies information about normal vibrations coupled to the electronic excitation.
4. The anisotropy of light absorption or emission gives information about the orientation of the electronic transitions and is very susceptible to changes of molecular geometry and symmetry.

Electronic excitation spectra in the UV and visible regions can also supply extremely valuable information about molecular structure (see Sect. 4.3).

References

1. Kortüm G (1962) Kolorimetrie, Photometrie und Spektrometrie, Kap 1.5, 4. Aufl. Springer Berlin Göttingen Heidelberg, S 21 ff
2. Perkampus H-H, Sandemann I, Timmons CJ (Hrsg) (1966–1971) DMS-UV-Atlas, Vol I–V. Verlag Chemie, Butterworth, London Weinheim
3. Kortüm G (1936) Z Physik Chemie (B) 33:243
4. Hellma-Küvetten, Mülheim/Baden, Katalog 67/32 u. 76/34
5. Uvasole, Lösungsmittel und Substanzen für die Spektroskopie. Merck, Darmstadt
6. Baker Analyzed Reagenz für die UV-Spektroskopie, Katalog 780, Baker-Chemikalien, Groß-Gerau
7. McClure DS, Blake NW, Hanst PL (1954) J Chem Phys 22:255; McGlynn SP, Azumi T, Hasha M (1964) J Chem Phys 40:507; McGlynn SP (1958) Chem Rev 58:1113; Evans DF: J Chem Soc 1957:1351; 1959:2753; Robinson GW (1961) J Mol Spectrosc 6:58
8. Birks JB (1973) In: Organic Molecular Photophysics, Chap 1. Vol 1. Wiley, London New York Sidney Toronto, p 1 ff
9. Perkampus H-H, Vollbrecht HR (1971) Spectrochim Acta Part A 27a:2173
10. Jaffé HH, Orchin M (1962) Theory and Applications of Ultraviolet Spectroscopy. Wiley, New York London
11. Murrell JN (1963) The Theory of the Electronic Spectra of Organic Molecules. Methuen, London
12. Becker RS (1969) Theory and Interpretation of Fluorescence and Phosphorescence. Wiley, New York London Sidney Toronto

3 Photometers and Spectrophotometers

The basic principles of the construction of photometers and spectrophotometers is the same; i.e. they consist of a light source, monochromator or filter, cuvette compartment, detector and amplifier with an indicating device.

Table 1. Filter combinations for isolating emission lines from metal-vapor discharge lamps according to [1]. The figures in the column 'filter combinations' refer to Tables 2 and 3

Element	λ in nm	Filter combinations Filter-No.	Percentage transmission at room temp. (approximate)	For surpressing the near IR and residual red radiation Filter-No.
Zn	308	4+32+33	5	16 (36)*
Hg	313	4+34	35	16 (36)
Cd	326	4+32+34	5	16 (36)
Hg	334	4+32+35	10	16 (36)
Zn	328/30/35	4+32+35	2	16 (36)
Tl	352/53	2+10+32	10	16 (36)
Hg	365	5+9+31	20	16 (36)
Tl	378	2+22	30	16 (36)
Hg	404/07	1+3+20+9	1	16 (36)
Hg	435/36	10+17+6	4	16 (36)
Cs	456/59	9+22	40	16 (36)
Cd	468/80	9+18	25	16 (36)
Zn	468/72/81	9+18	25	16 (36)
Cd	509	7+21+8	20	16 (36)
Tl	535	14+19	35	16 (36)
Hg	546	15+23+13+8	10	16 (36)
Hg	577/79	12+24+12	15	16 (36)
He	588	12+25+12	10	16 (36)
Na	589	12+25+12	10	16 (36)
Zn	636	26	85	
Ne	638–668	26	90	
Cd	644	26	90	
He	668	27+11	20	
He	707	29	65	
K	767/70	30+29	25	
Rb	780/95	30+29	25	
Cs	794–921	30+29	10	
Cs	852–921	30+28	1	

* Liquid filter No. 36 (see Table 3) can be used instead of NIR-filter 16.

3.1 Photometers

In *photometers*, also called non-dispersive instruments, the monochromator is replaced by a set of filters by means of which specific spectral ranges can be selected from the continuum of a light source, e.g. a tungsten-halogen lamp in the visible region. Mercury-vapor high pressure lamps combined with interference filters are often used so that mercury lines at 334, 365, 404/407, 435/436, 546 and 577/579 nm can be utilized. Photometers having a line source as the light source and fitted with an interference filter provide a more monochromatic beam. The number of spectral lines can be extended by using other metal-vapor discharge lamps; for example, a cad-

Table 2. Glass filters in common use, from [2]

Filter No.	Code No.	Thickness mm	Filter No.	Code No.	Thickness mm
1	UG2	1	17	GG435	4
2	UG3	2	18	GG455	3
3	UG3	2	19	GG475	2
4	UG5	3	20	GG385	5
5	UG11	2	21	GG495	2
6	BG3	2	22	GG375	2
7	BG7	1	23	OG530	1
8	BG8	2	24	OG570	3
9	BG12	2	25	OG590	2
10	BG12	4	26	RG610	2
11	KG3	3	27	RG665	2
12	BG18	2	28	RG1000	2
13	BG18	3	29	RGN9	2
14	BG18	5	30	KG1	2
15	BG20	5	31	WG360	1
16	BG38	3			

Table 3. The most generally useful liquid filters

Filter No.	Description	Quantity/l H_2O	Pathlength/mm (inside dimension of the cuvette)
32	Nickel-cobalt sulphate $NiSO_4 + CoSO_4$	303 g + 86.5 g	20
33	Picric acid	16 mg	20
34	Potassium chromate	150 mg	20
35	Nitric acid	N/5	20
36	Copper sulphate $CuSO_4 + 5H_2O$	57 g	10

mium lamp supplies lines at 326, 468/480, 509 and 644 nm. A summary is given in Table 1 [1]. Details of the filters are given in Tables 2 and 3 [2]; see also [3].

Photometers are used in the photometric determination of single substances (see below). In recent years, they have also been employed in the clinical-chemical and biochemical fields, as well as in special equipment developed for the analysis of waste gases.

3.2 Spectrophotometers

In a spectrophotometer, nowadays more usually termed a spectrometer, the measuring light is split up (dispersed) into its constituent wavelengths by a prism or grating monochromator. With a deuterium lamp for the UV region and a tungsten (tungsten-halogen) lamp for the VIS region, these instruments allow the continuous variation of the measurement wavelength over the whole spectral region. They are also called dispersive spectrometers. Most instruments cover the range 190 to 900 nm [4].

We differentiate between *single-beam* and *double-beam* instruments.

Single-beam instruments generally operate on the substitution principle, i.e. the reference and measurement cuvette are placed one after another in the path of the light. The 100% point, previously set manually via the slit or by changing the amplification, is today adjusted automatically in most instruments which normally display the spectrophotometric result as percentage transmittance or absorbance in digital form.

In a double-beam instrument, the primary light beam is split and directed along two paths which traverse alternately the reference and measurement cuvette, which are approximately 10–15 cm apart. Thus, after both beams have been refocused, light of varying intensity falls onto the detector and generates an alternating-voltage signal. This principle forms the basis of recording spectrophotometers.

In the case of fixed beam-splitting elements, the alternating direction of the light through the two cuvettes must be made by means of a chopper. In the case of a rotating sector mirror, the beam-splitting element itself controls this function. Figure 5 shows the optical system of a double-beam instrument with a double monochromator (Perkin-Elmer Lambda 9).

The monochromator is the most important component of a spectrophotometer. Here we must distinguish between instruments with *single monochromators* and instruments with *double monochromators*. A double monochromator has the important advantage that the proportion of stray light is very small. Stray light is light from another spectral region which is superimposed on the 'useful light' of the spectral region which is selected for measurement. It can distort the required measurement considerably (see Sect. 3.3).

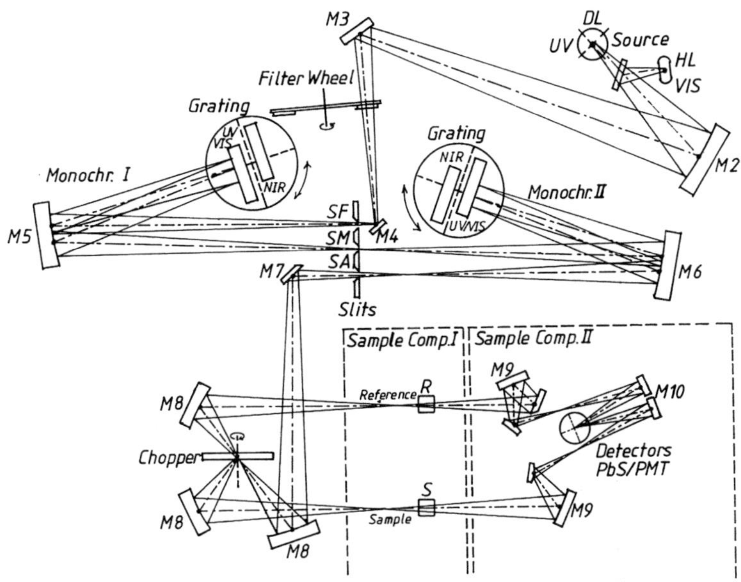

Fig. 5. Optical layout of an UV-VIS-NIR spectrometer with a double monochromator (Perkin-Elmer Lambda 9) M_1, M_3, M_4, M_7 plane mirrors, M_2, M_5, M_6, M_8, M_9, M_{10} toroidal mirrors, SE entry slit monochromator I, SM centre slit = entry slit monochromator II, SA exit slit monochromator II. Detector: UV-VIS region photomultiplier NIR region PbS-detector holographic gratings for UV-VIS and NIR on a turntable

Monochromators in use today are almost exclusively grating monochromators and, depending on the quality of the gratings employed, the proportion of scattered light lies between 0.05 and 0.005% for single-grating monochromators. The equipment manufacturers have introduced the notion of the "proportion of scattered light" as an index of quality. This term refers to the intensity of the light leaving the exit slit which has a wavelength in the immediate neighborhood of that of the desired light, λ_0. Erroneously, it is also sometimes called "stray light", see Sect. 3.3. For double monochromators, the proportion of scattered light is lower by about two powers of ten. These figures are average values obtained from the manufacturer's literature. Holographic gratings provide a substantial improvement in stray light characteristics over ruled ones. See Sect. 3.3 for the elimination or determination of stray light.

The advantage of a grating vis-a-vis a prism lies in the fact that a grating shows a dispersion which is linear with wavelength. The correlation between the resolving power in wavenumbers $\tilde{\nu}$ and wavelength λ in nm for the spectral bandwidth $\Delta\lambda = 1$ nm is shown in Table 4. For any other value of $\Delta\lambda$, $\Delta\tilde{\nu}$ is obtained by multiplying values from this table with the appropriate $\Delta\lambda$. It can be seen that the resolving power for a constant spectral bandwidth increases from the UV to the visible. Alternatively, for a constant resolving power the bandwidth decreases from UV to visible.

Table 4. Resolving power $\Delta\tilde{\nu}$ in cm^{-1} in the UV-VIS region for a spectral bandwidth of $\Delta\lambda = 1$ nm

λ [nm]	$\tilde{\nu}$ [cm^{-1}]	$\Delta\tilde{\nu}$ [cm^{-1}]
200	50000	250
300	33333	110
400	25000	62
500	20000	40
600	16666	28
700	14286	20
800	12500	16

With a few exceptions, all modern instruments are fitted with grating monochromators.

With fully automated instruments, depending on the sophistication of the software, the following functions can be recalled or continuously monitored by a microcomputer:

Base-line correction; conversion of analog to digital data; recorder, printer or plotter control including format selection with graph plotters; conversion of extinction values to concentrations; input of the recording range; recording in wavelength or wavenumber; repeat-recording over selected wavelength ranges or at different wavelength and time intervals; lamp and filter changes; formation of the 1st and 2nd and, if necessary, of higher derivatives; generation of Good Laboratory Practice (GLP) protocols by the printer, e.g. printout of analysis data with sample identification for sets of measurements; calculation of difference spectra.

The scope of application can be considerably extended by fitting supplementary modules and accessories, such as thermostated or temperature-ramped cuvette changers controlled by a microprocessor, fluorescence attachments, accessories for diffuse and specular reflection spectroscopy, and for enzyme kinetics, sipper systems for repeated measurements, gel scanners and chromatographic attachments.

Several newer techniques such as *derivative spectroscopy* [5], see Sect. 5.1, and *dual-* or *double-wavelength spectroscopy* [6], see Sect. 5.2 have recently gained in importance.

Generally, instruments fitted with microcomputers allow the recording of *1st and 2nd order derivative spectra.* This method has come increasingly to the fore in analytical applications since it can improve the sensitivity of detection considerably.

Whilst devices for derivative spectroscopy can be fitted to many recording spectrophotometers post manufacture, true double-wavelength spectroscopy requires a special instrument whose most important components are two optically identical monochromators.

However, double-wavelength spectroscopy can also be pursued with a microcomputer-controlled spectrophotometer by entering the extinction value at a specific wavelength λ_1 into the memory and by retrieving it from the memory for comparison with the extinction values at wavelengths λ_2, λ_3.... Several manufacturers have taken advantage of this possibility in their software.

Diode-array spectrometers are an interesting new development pioneered by Hewlet-Packard. Instead of a monochromator, a *polychromator* is used in which the dispersed light of a continuum source is brought to a focus in one plane, the focal plane of the instrument. This method of operation corresponds to the classical method where a photographic plate, which "recorded" the whole spectrum directly, was fitted in the focal plane of a spectrograph. Today a silicon-diode array detector is used, instead of a photographic plate, and this allows direct, fast electronic processing of the spectral information stored for a short time in one row of 256 or 512 diodes (channels). The components are not arranged as in a conventional spectrophotometer, in particular the cuvettes are mounted between the light source and the entry slit of the polychromator.

Figure 6 shows the optical layout of the Perkin-Elmer Lambda-Array-UV-VIS Spectrometer 3840. The light of a deuterium lamp is focused on to the sample by the plane (flat) and toroidal mirrors (M_2 and M_3) and, after traversing the sample, arrives via mirror M_4 and another plane mirror at the entry slit of the polychromator. Here the continuum of the light source is directed onto a grating via M_5. Finally the light, now spectrally dispersed, falls via mirror M_6 onto the photodiode array (PDA) detector with 512 photodiodes. Two gratings are mounted on a turntable. The first grating (100 lines/mm) is used for the fast acquisition of a survey spectrum in the 200–900 nm region (theoretical resolution 1.5 nm, effective resolution ca. 4.5 nm, i.e. the survey mode). The second grating (600 lines/mm) is used in a high-performance mode in which it is rotated in 7 steps to achieve a high resolution ($\approx$0.25 or $\approx$0.75 nm). In this mode, spectral regions of ca. 100 nm arrive at the diode array, i.e. the whole region of measurement is divided into 7 sections. Depending on the width of the selected λ region, the time for measuring a whole spectrum is approximately 8–16 s. The deuterium lamp (type D 802 or D 902, Heraeus Original Hanau) transmits light along the optical axis and, as shown in Fig. 6, this allows the introduction of the beam of a tungsten lamp into the light path. The W and source filter wheels are used to cut out one lamp when measuring with the other. All functions (filter change, shutter, rotation of grating, PDA read-out, collection of data, etc.) are controlled by the Perkin-Elmer PC 7500.

Hewlett-Packard have developed a similar instrument, model HP 8452 A. Philips and Milton Roy also market diode array spectrometers.

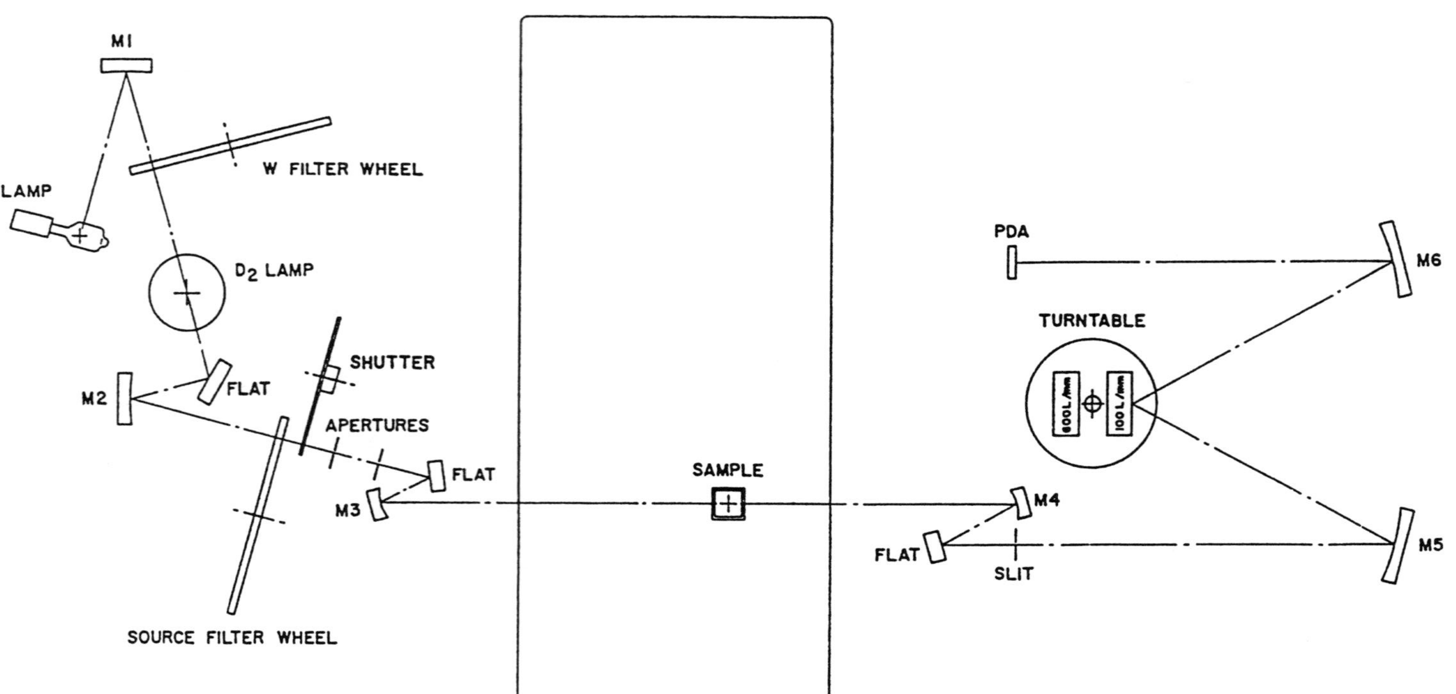

Fig. 6. Lambda array 3840 optical layout

3.3 The Stray Light Error

3.3.1 General Observations

Since the vast majority of UV-VIS spectrophotometers are still fitted with a single monochromator, errors can occur due to monochromator stray light at the limits of the monochromator transmission (e.g. $\lambda \leqslant 220$ nm $\hat{=} \geqslant 45000\ \mathrm{cm}^{-1}$); particularly in the case of small transmission values.

By stray light we mean light of other wavelengths which is superimposed upon the useful light. Often mistakenly called scattered light, it is caused by scattering at the optical surfaces in the monochromator.

If a monochromator is set to wavelength λ_0 and if the slitwidth corresponds to the effective bandwidth $\Delta\lambda$, the useful light lies in the region between

$$\lambda_0 - \Delta\lambda \quad \text{and} \quad \lambda_0 + \Delta\lambda \ . \tag{8}$$

We shall call this the *useful-light region*. In ideal circumstances, a monochromator should only transmit in the useful light region with the transmission decreasing linearly from λ_0 in both directions. However, due to stray light there is a certain transmission by the monochromator of light outside the useful light region. Although its intensity is small (order of magnitude 10^{-5}) it can still be significant since the radiation detector sums the stray light over the whole wavelength region to which it is sensitive and in which the light source emits radiation.

The proportion of the photoelectric current given out by the detector and reaching the display, which is due to stray light, is the critical measure of the stray light effect when making spectrophotometric measurements.

For that reason, we understand by the term *proportion of stray light* the ratio of the photoelectric current arising from stray light to the total photoelectric current. Although the proportion of the photoelectric current arising from the stray light is generally low ($<0.1\%$), it can cause significant problems if the photoelectric current due to the useful light becomes relatively small. In practice, this occurs in the following cases:

1. The useful light can be weakened by absorption occurring in the light path whilst the stray light is hardly diminished. This occurs particularly below 230 nm because:

   a) The optical elements in the light path (envelope of the hydrogen lamp, source mirror, lenses, mirrors, prisms or gratings in the monochromator, lenses in the sample changer, multiplier envelope) absorb increasingly with decreasing wavelength. In addition, there is the effect of the presence of absorbing deposits on the accessible optical surfaces (envelope of the hydrogen lamp, lenses in the sample changer, cuvettes). Furthermore, atmospheric oxygen which is present in the optical path of the instrument absorbs at wavelengths below 200 nm.

b) If the solvent absorbs in the short-wavelength UV region but the long-wavelength stray light is transmitted undiminished, the proportion of stray light therefore increases. Special attention should be paid to this possibility since most solvents absorb in the short-wavelength UV region due to the presence of impurities if they have not been especially purified. For many solvents, preferred because of their solvent properties, the transmission at $40000\,cm^{-1}$ (= 250 nm) is practically zero, therefore, for these solvents, stray light must be taken into consideration below 260 to 270 nm.

2. In some spectral ranges, the intensity of the radiation from a light source in the useful-light region is relatively small in relation to the intensity in the region generating the stray light. This is the case when measuring with an incandescent lamp in the range between 320 and 400 nm. Therefore, stray-light protective filters, which absorb the long-wavelength stray light, should be used in this region. These filters can reduce the proportion of stray light to less than 0.2% for measurements made with solvents free from the problems mentioned under 1 b.
3. In certain spectral ranges, the detector sensitivity in the useful-light region is relatively small compared with that in the stray light region. This is the case at the long-wavelength limit of the detector sensitivity, i.e. for photomultipliers above approximately 620 nm and for photocells above 1.1 µm. For that reason, measurements cannot be extended beyond these limits without the risk of a considerable error unless the short-wavelength stray light is reduced by special filters.

Case 1 is particularly important in practice and requires careful control of the proportion of stray light during measurement if the solvent shows considerable absorption in the useful-light region. In any event, the absorbance of the solvent should be tested when measuring below 230 nm. For this purpose, the absorbance of the reference cuvette can be measured against air. If the absorbance of the solvent is greater than approximately 0.5, the possibility of reducing the absorbance either by purifying the solvent or by reducing the pathlength should be investigated.

A clear indication of a stray light error by absorbing solvents manifests itself as follows. If the measurement is made with different pathlengths and the extinction coefficient recorded as a function of the wavelength, the same value of the extinction coefficient should be obtained for all pathlengths. This is often the case, within the error threshold, for wavelengths above ca. 250 nm. In contrast, in the presence of a stray light error, the curves diverge with decreasing wavelength in the sense that smaller values of the extinction coefficient are obtained for longer pathlengths. Obviously, values obtained at the shortest pathlength are the most reliable ones.

3.3.2 The Stray Light Error of Transmission and Absorbance and Its Measurement

The useful light leaving the reference cuvette induces the photoelectric current I_0 in the detector. The photoelectric current generated by the useful light leaving the sample cuvette is I and the real transmission (T) of the sample is

$$T = \frac{I}{I_0} . \tag{9}$$

The stray light induces the additional photoelectric current I_f. Consequently, the measurement gives a false transmission T′ which has the value

$$T' = \frac{I+I_f}{I_0+I_f} . \tag{10}$$

This assumes, in the first instance, that the stray light is reduced by the sample in the same manner as by the solvent. This assumption is adequate in many cases.

On introducing the proportion of stray light, p,

$$p = \frac{I_f}{I_0+I_f}$$

we obtain

$$T' = T(1-p)+p . \tag{11}$$

If the proportion of stray light is known, the true transmission can be calculated from the value distorted by the stray light as

$$T = \frac{T'-p}{1-p} . \tag{12}$$

When moving from transmission to absorbance the following equations

$$A = -\log T;\ A' = -\log T'$$

provide the absorbance error caused by stray light

$$\Delta A = A'-A = \log T-\log[T(1-p)+p] . \tag{13}$$

The *relative error* $\Delta A'$ of the absorbance is of particular interest in spectrophotometry. This value is plotted in Fig. 7 as a function of the distorted absorbance A′ which is read directly [7]. The proportion of stray light, p, has been included as a parameter.

The graph shows that the relative error in the absorbance increases rapidly with the value of the absorbance for a given proportion of stray light p. Therefore, the pathlength and concentration must be selected for all measurements such that the absorbance is not too large. At p values of the order

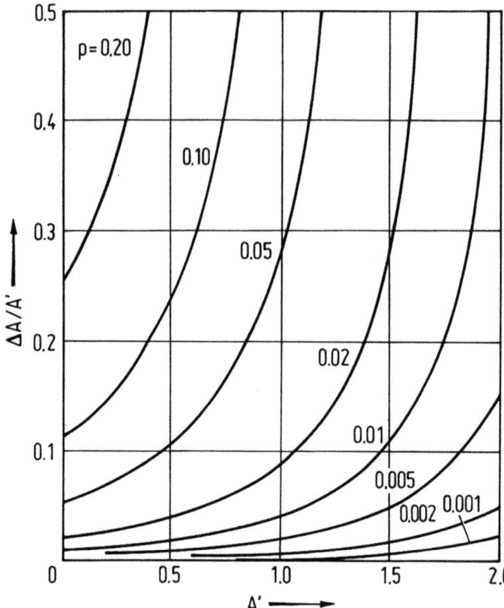

Fig. 7. Effect of stray light on absorbance A

of magnitude of $0.005 \leqslant p \leqslant 0.02$, an absorbance of 0.6 to 0.8 should not be exceeded. The graph shows for each individual case the greatest value of absorbance which may be used in order to avoid exceeding a specific relative error in the measurement.

The useful light must be removed from the light path when measuring the proportion of stray light. Then the photoelectric current I_f caused solely by stray light can be measured. The useful light can be removed by absorption or shielding [8].

After the proportion of stray light, p, has been established, the correct transmission (T) is given by:

$$T = \frac{T' - p}{1 - p} .$$

In practice, to determine the true transmittance, T in the presence of stray light, p, we can proceed as follows. Cuvettes containing solvent (L), sample solution (P) and a very strongly absorbing solution (K) are used. The cuvette K has the same pathlength and contains the substance to be measured at such a concentration that the actual transmission lies below 0.1% in the useful-light region. We then measure P against L (result T'), and K against L (result p) and calculate the corrected transmission T of the sample, using the above equation.

It follows from the definitions of T, T' and p that we must use the values as a fraction of 1.0 and not as percentage figures in formulae, i.e. $T' = 0.32$ (instead of 32%) and $p = 0.01$ (instead of 1%).

This method of approximation, when determining p, is adequately accurate if the useful-light region coincides with a strong absorption band since, in this event, the test solution K will not weaken the stray light much more than the sample P or solvent L. However, correction of the stray light error is particularly important in this case because the sample is likely to have a relatively high absorption in the useful-light region.

Under less favorable conditions, where the sample P, when at sufficiently high concentration for use as test sample K, absorbs the stray light considerably more than the actual sample P, another substance must be used as test sample. A suitable substance can easily be found when measuring in the short-wavelength UV region. The assumption that L and P always absorb the stray light to the same extent should be avoided. In this case the following equation applies

$$T' = \frac{I + I_f}{I_0 + I_{f_0}} ,$$

but the stray-light photoelectric currents, I_f and I_{f_0}, differ from one another. It is obvious from this that two values, p′ and p″, can be measured with an ideal test sample (p′) or by cutting out the useful light (p″).

$$p' = \frac{I_f}{I_0 + I_{f_0}} ,$$

$$p'' = \frac{I_{f_0}}{I_0 + I_{f_0}} .$$

In this case, the following applies for the true transmission

$$T = \frac{T' - p'}{1 - p''} .$$

Whether it is appropriate to use this improved approximation to correct T′ depends upon how accurately the two stray light ratios p′ and p″ can be measured.

The work of Luther et al. [9] and Luck [10] should be consulted with regard to the *stray-light error of UV measurements*. Further details and procedures for and evaluation of the stray-light error are given by Burgess [11], Cook et al. [12], Poulson [13], Renle [14] and Kaye [15].

Reference may be made to Derkosch and Gauglitz for further details of measurement techniques [16, 17].

3.4 Light Sources for UV-VIS Spectroscopy

The light source is one of the most important components of a UV-VIS spectrometer; and it must be a continuum source in order to measure a complete absorption spectrum in the UV-VIS region of 190–900 nm.

In the visible spectral region, this requirement is met by the *tungsten lamp* which is a black-body source, the spectral energy distribution of which is described by Planck's radiation formula [18]. According to Wien's law, the energy maximum of the energy distribution lying in the NIR region moves to shorter wavelengths with increasing temperature [19]. The short-wavelength edge of the distribution in the visible region is thereby raised, hence the radiation yield increases in this region. However, the higher temperature of a tungsten coil means a shorter lamp life since tungsten evaporates from the coil. The evaporated tungsten precipitates on the cooler glass envelope. The deposit on the glass envelope then causes a reduction of radiated energy due to absorption of the emitted radiation, see Sect. 3.3.1.

This problem is overcome in modern *halogen lamps*. The added halogen (e.g. iodine vapor) forms a volatile compound with the evaporated tungsten which decomposes on the hot tungsten coil. Therefore, even at higher temperatures the tungsten coil thickness remains almost constant even during prolonged operation of the lamp. At the same time, the condensation of evaporated tungsten on the bulb is considerably reduced. Thus, both effects result in a higher radiation output in the visible spectral region and a longer lamp life.

As with any other black-body source, the spectral energy distribution of a tungsten lamp decreases rapidly below 400 nm. Therefore, these lamps cannot be employed in the UV region. In the region below 350 nm gas-discharge lamps are therefore used as radiation sources; of these the hydrogen or deuterium lamp is the most important source of light. The hydrogen discharge provides a continuous spectrum between 160 and 400 nm (62500–25000 cm^{-1}). For this reason, below ca. 350 nm (above 28500 cm^{-1}), hydrogen or deuterium lamps are commonly used as light sources in spectrophotometers. Their construction is described briefly in the following [20]:

A tungsten coil is fitted as anode on the axis of a cylindrical, thin-walled, quartz discharge tube. An activated tungsten double coil mounted laterally is generally used as cathode. The lamp is filled with either hydrogen or deuterium at ca. 10 Torr. A heating voltage is applied to the cathode coil so that the ignition potential remains low (200–400 V) because of the favorable thermal emission properties. Hydrogen, the lightest gas, has a very high diffusion velocity; consequently, energy losses due to thermal conduction are very large and the radiation gain is correspondingly low. For that reason, deuterium is used in modern lamps which improves the radiation yield by about 30% as a result of the doubling of the molecular weight. In order to produce the highest possible radiation intensities it is necessary to restrict the discharge between cathode and anode by means of an aperture of small cross-section (ca. 1 mm^2) formed from a high-melting point metal (molybdenum).

Between 160 and ca. 400 nm, an emission occurs due to the transition from a stable excited state, which is the lowest triplet state, ${}^3\Sigma_g^+$, of the H_2

molecule, to the repulsive state $^3\Sigma_u^+$ [21]. Thus, the resulting spectrum is continuous. Dissociation occurs after the system has reverted to the lower state and, for that reason, the continuous spectrum emitted is also referred to as dissociation radiation [21]. The hydrogen or deuterium atoms so formed recombine to molecules H_2 or D_2 on the cold surfaces in the discharge chamber. In addition to the continuum, the Balmer series of the H or D atoms can be seen in the visible region and the appropriate H_α-, D_α-($n = 3 \rightarrow n = 2$) or H_β-($n = 4 \rightarrow n = 2$) lines can be used for calibrating the wavelength scale of a UV-VIS spectrometer. If the lamps are filled with a hydrogen-deuterium mixture the Balmer series of both gases are emitted. The H_β-line lies at 486.12 nm and the D_β-line at 485.99 nm, and the difference of $\Delta\lambda = 0.13$ nm or $\Delta\tilde{\nu} = 5.5$ cm^{-1} can be conveniently used to check the resolving power of a UV-VIS spectrometer. For manufacturers of deuterium lamps see [22].

In addition to these two most important lamps, others are employed in special applications where, for example, a higher radiation output in the VIS and UV region is of particular interest. Noble-gas discharge lamps show a prominent continuum, and of these, the *xenon-high pressure lamp* is the most commonly used. The continuum of a xenon discharge corresponds to recombination radiation, i.e. xenon atoms ionized at high gas temperatures recombine with electrons formed during ionization and emit radiation in the UV-VIS and NIR region.

Manufacturers produce xenon lamps with powers from 75 W to several kW [23], and they are preferred in fluorescence and luminescence excitation spectroscopy (see Sect. 5.5) as well as in photoacoustic spectroscopy (see Sect. 5.4). In the UV-VIS region, these lamps are also used as light sources in microscope spectrometers. For these applications xenon lamps of up to 450 W are usually employed.

The quartz bulb of a xenon-maximum pressure lamp is considerably thicker than that of a deuterium lamp for reasons of safety. Thus, the intensity of UV radiation which is already diminishing toward shorter wavelengths is increasingly absorbed by the quartz below 250 nm. For this reason, xenon lamps are produced for special applications with bulbs made of a quartz material such as Spectrosil which is highly transparent in the UV region [23].

In contrast to metal-vapor discharge lamps (e.g. mercury lamps), xenon lamps require no warming-up time because they reach their full power immediately after switching on. Hence, they can be easily modulated or used as flashlamps.

In addition to these continuous light sources, metal-vapor discharge lamps are often employed. The most important lamp of this type is the *mercury vapor lamp* which is produced for low pressure, high pressure or maximum pressure [24]. As line sources, they are specifically used for isolating definite spectral lines by means of suitable interference filters or filter combinations (cf. Table 1, page 10), and therefore, they are employed in photometers as light sources.

The mercury low-pressure lamp, operated at a pressure of 0.006 Torr, emits almost exclusively an intercombination line at $\lambda = 253.7$ nm and is suitable for use in photochemistry. Depending on its power, a mercury high pressure lamp operates at a pressure of between 10 to 50 atm. In addition to the emission of the characteristic mercury lines, the emission spectrum of a high pressure lamp also shows a continuous background. On account of the high vapor pressure in the discharge, the spectral lines are considerably broadened. A mercury medium-pressure lamp has a considerably lower operating pressure and the spectral lines are relatively clear. Therefore, the use of this lamp is preferred in the UV-VIS region and as an excitation light source in photochemistry.

Schäfer and Heinrich have given a detailed account of the light sources briefly described here [25].

In recent years, lasers have come to the fore as light sources for special applications; and noble-gas and dye lasers in particular should be mentioned. The dye laser has the advantage of being completely tunable over a wavelength range of ca. 600–1000 nm depending on the dye used. As a rule, an argon-ion laser is used as a pump light source for a dye laser. W. Demtröder has given a detailed account of the applications of lasers in spectroscopy [26].

References

1. Osram, Druckschrift (1978) Licht für Kinoprojektion, Technik und Wissenschaft, Ausg Dez, 869:1
2. Jenaer Glaswerke Schott & Gen, Mainz, Farb- und Filterglas
3. Perkampus, H-H (1980) In: Ullmanns Encyklopädie der technischen Chemie, 4. Aufl, Bd 5. Verlag Chemie, Weinheim, 269 ff
4. Perkampus, H-H (1983) In: Analytiker-Taschenbuch, Bd 3. Springer, Berlin Heidelberg New York, S 279–316
5. Talsky G, Mayring L, Kreuzer H (1978) Angew Chem 90:840
6. Shibata S (1976) Angew Chem 88:750
7. Hogness TR, Zscheile Jr FP, Sidwell Jr AE (1937) J Phys Chem 41:379
8. Preston IS (1936) J Scient Instr 13:3681
9. Luther H, Pokkels G (1955) Z Elektrochem Ber Bunsenges 59:159
10. Luck W (1960) ibid 64:676
11. Burgess C, Knowles A (1981) Standards in Absorption Spectrometry, Ultraviolet Spectrometry Group, Vol I. Chapman and Hall
12. Cook RB, Jankow AR (1972) J Chem Ed 49:405
13. Poulson RE (1964) Appl Opt 3:99
14. Renle A (1971) Coll Spectr Int XVI 1:107
15. Kaye W (1981) Anal Chem 53:2201
16. Derkosch J (1967) Absorptionsspektralanalyse im ultravioletten, sichtbaren und infraroten Gebiet, Bd 5 der Methoden der Analyse in der Chemie. Akad Verlagsges, Frankfurt/M
17. Gauglitz G (1983) Prakt Spektroskopie. Attempto, Tübingen
18. Planck M (1901) Ann Phys 4:553
19. Wien W (1894) Ann Phys, series 2, 52:132

20. Kiefer J (ed) (1977) Ultraviolette Strahlen. Walter de Gruyter, Berlin, p 96
21. Herzberg G (1967) Molecular spectra and molecular structure. I Diatomic molecules
22. Heraeus WC, Brochure D 310686/2C 7.86 VNKo. Produktbereich Original Hanau, Postfach 1553, D-6450 Hanau 1
23. Osram GmbH, Berlin-Munich, Brochure "Licht für Kinoprojektion, Technik und Wissenschaft". Issue August 1988/FO 302
24. Heraeus, Brochure D 310218/1C, 10.85 VNKo; D 310531/1C 10.85; D 310311/1C 10.85. Produktbereich Original Hanau, see 22
25. Schäfer V, Heinrich G (1977) In: Kiefer J (ed) Ultraviolette Strahlen, Chap 3. Walter de Gruyter, Berlin
26. Demtröder W (1988) Laser Spectroscopy. Springer, Berlin Heidelberg New York, 3rd printing with corrections

4 Analytical Applications of UV-VIS Spectroscopy

For about 130 years, the Bouguer-Lambert-Beer law has been used as the quantitative basis of absorption spectroscopy. Bouguer established empirically a correlation between pathlength and light absorption in 1729. This correlation was then formulated mathematically by Lambert in 1760 [1] and Beer discovered the dependence upon the concentration in 1852 [2]. Initially, the human eye was the detector for comparing different light intensities. In 1925, Pulfrich [3] introduced his photometer as "a photometer appropriately adapted to exceed the levels of sensitivity of the human eye, called a step photometer...". Our eyes are capable of assessing the uniformity of two light densities with an accuracy of approximately 1%. The principle of colorimetry or visual photometry is based on this fact (see [4–6] etc.).

The Bouguer-Lambert-Beer law applies to a *dilute* solution containing one component or equally to a dilute solution of *several* components. The measured absorbance, A_1, of multicomponent solutions at wavenumber $\tilde{\nu}_1$ then equals the sum of absorbances which we would obtain separately for each solution had we measured the solutions individually at the same wavenumber $\tilde{\nu}_1$ or wavelength λ_1 and pathlength d:

$$A_1 = A_{11} + A_{12} + A_{13} + \ldots = \sum_j A_{1j} \,. \tag{14}$$

We can then write the general formula:

$$A_i = \varepsilon_{i1} \cdot c_i \cdot d + \varepsilon_{i2} c_2 d + \varepsilon_{i3} \cdot c_3 \cdot d + \ldots = d \sum_{j=1}^{n} \varepsilon_{ij} \cdot c_i = \sum_{j=1}^{n} A_{ij} \,. \tag{14a}$$

Index i refers to wavelength λ or wavenumber $\tilde{\nu}$ and the second index j to the components.

Equation (4) is the basic equation for the photometric determination of a single substance and Eq. (14a) is the foundation for photometric multicomponent analysis.

4.1 Photometric Determination of a Single Substance

Equation (4) shows the simple linear correlation between absorbance A and concentration c and the molar decadic extinction coefficient, $\varepsilon_{\tilde{\nu}}$.

The concentration c

$$c = \frac{A_{\tilde{\nu}}}{\varepsilon_{\tilde{\nu}} \cdot d} \tag{15}$$

can be determined if the extinction coefficient, $\varepsilon_{\tilde{\nu}}$, of the substance to be determined is known.

Equation (15) shows the concentation ranges that can be determined. At a maximum extinction coefficient of $\varepsilon_{\tilde{\nu}} = 10^5 \, l \, mol^{-1} \, cm^{-1}$,

a pathlength of $d = 1$ cm and
a lower absorbance value of $A_{\tilde{\nu}} = 0.1$

we obtain a detectable concentration at $c = 10^{-6} \, mol \, l^{-1}$.

In modern equipment, the microcomputer ensures a base-line stability of $10^{-3}-10^{-4}$ absorbance units and a noise of the same or smaller order of magnitude. Thus, absorbance values of 0.01 to 0.001 can be measured so that the limit of detection lies at $c = 10^{-7} \, mol \, l^{-1}$ and, in the most favorable case, for an appropriately selected pathlength, at $c = 10^{-8} \, mol \, l^{-1}$.

However, extinction coefficients in the given range of $10^5 \, l \, mol^{-1} \, cm^{-1}$ are generally the exception. Most compounds absorbing in the UV-VIS region have considerably smaller extinction coefficients which lie in the region of

$$10^3 \leqslant \varepsilon \leqslant 5 \times 10^4 \, l \, mol^{-1} \, cm^{-1} \; .$$

This range of extinction coefficients applies to many organic compounds with chromophoric systems. However, typical dyes or dye-like chromophoric systems can have ε values of approximately $100000 \, l \, mol^{-1} \, cm^{-1}$.

In contrast, the visible colors of metal cations, which we are frequently required to analyze, are characterized by small values of the extinction coefficients. The situation is more favorable with strongly colored complex compounds. The spectrum of hexaaquo-copper(II)-perchlorate in 0.1 N $HClO_4$ (Fig. 8a) has an absorption maximum at $\tilde{\nu} = 12400 \, cm^{-1}$ with $\varepsilon = 11 \, l \, mol^{-1} \, cm^{-1}$ [7]. When complexed with 8-hydroxyquinoline and 1-hydroxyacridine (Figs. 8b, c) absorption bands occur in the visible region with extinction coefficients [8] of

$\varepsilon = 2.8 \times 10^3 \, l \, mol^{-1} \, cm^{-1}$ for the 8-hydroxyquinoline complex in ethanol

and

$\varepsilon = 2.7 \times 10^3 \, l \, mol^{-1} \, cm^{-1}$ for the 8-hydroxyacridine complex in chloroform.

Since the extinction coefficients of such complexes are related to the molecular weights of the complexes, the sensitivity of the photometric method

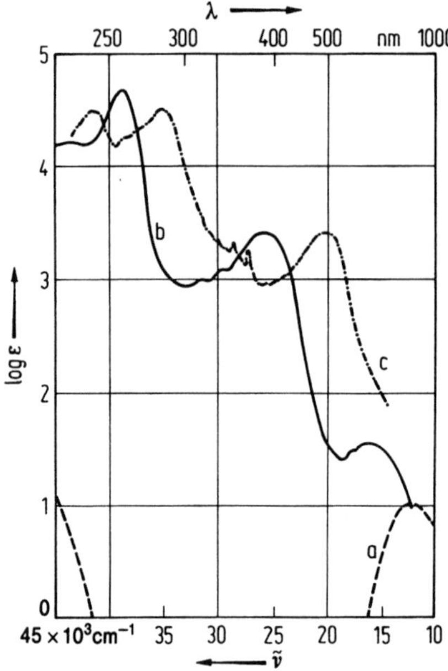

Fig. 8. Absorption spectrum of *a* hexaaquo-copper(II) in H_2O, *b* copper(II)-8-hydroxy-quinoline in ethanol, *c* copper(II)-4-hydroxyacridine in chloroform

for elements of a similar molecular weight can be compared for a given complexing agent. Ayres and Narang [9] introduced the *specific absorption* "a" for arriving at a generally applicable expression and hence a comparison of photometric methods. This absorption is given as

$$\mathrm{a} = \frac{\varepsilon_\lambda}{\text{atomic weight}} \times 10^{-3} . \qquad (16)$$

The molar decadic extinction coefficient ε_λ is adjusted for the atomic weight of the metal and this numerical value is multiplied by 10^{-3}. Consequently, the specific absorption "a" has the dimensions $\mathrm{ml\,g^{-1}\,cm^{-1}}$ and corresponds to the extinction of a solution which contains 1×10^{-6} g of the metal to be determined in 1 cm^3 at a 1 cm cuvette pathlength; this equals 1 ppm.

In addition to the specific absorption, Sandell [10] introduced the *sensitivity index* "S". It gives the number of micrograms of an element per ml in a solution which has absorbance A = 0.001 at a pathlength of 1 cm. The dimension is $10^{-6}\,\mathrm{g\,cm^{-2}}$ and S is related to the specific absorption "a" by

$$\mathrm{S} = \frac{10^{-3}}{\mathrm{a}} . \qquad (17)$$

4.1.1 Photometric Determination of Elements by Means of Complexing Agents

Metal cations which absorb weakly or not at all in the visible spectral region can be changed into strongly colored compounds by means of complexing agents. Strongly colored here means that absorption bands characteristic of the complex have an extinction coefficient $\varepsilon > 10^4\,\mathrm{l\,mol^{-1}\,cm^{-1}}$. Some complexing agents have proved to be extremely suitable, and these include dithizone [11] and 1-(2-pyridylazo)-2-naphthol (PAN) [12–16], 8-hydroxyquinoline (8-oxine) [17–20], formaldoxime [21–25], 1,10-phenanthroline and 2-2′dipyridyl [26–28], *N*-benzoylphenylhydroxylamine [29], morin [30, 31], Na-dithiocarbamate (DTC) [41–43] and the thiocyanate ion as an inorganic complexing agent [32–35] (see Fig. 9).

Since these complexing agents can form complexes with a large number of metals there is no great selectivity and in particular no specificity. This means that other metals can interfere with the metal we want to determine.

A *reagent* is referred to as *selective* if it reacts with a limited number of elements and *specific* if it undergoes the desired color reaction *with only one element* when we adhere to particular reaction conditions.

Among other things, the *selectivity* of a color reaction depends upon the chosen reagent, the oxidation number of the element and, in many cases, also upon the pH-value of the solution and above all upon the stability constant of the complex. This is extremely important since weaker complexes can be converted into more stable ones by recomplexing and therefore, they can no longer react with the first reagent. Thus, it is possible to mask interfering elements. This *masking technique* has stood the test of time and contributed to an extraordinarily improved selectivity and specificity of the method used for photometric determination. Marczenko [36] and Umland [37] have summarized the masking agents most commonly used today. Table 5 shows the elements and methods for their photometric determination for which detailed operating instructions are available.

In practice, we must proceed according to exactly defined conditions. This applies particularly if trace elements are to be analyzed. The operating instructions specific for each element and reagent have been summarized in excellent monographs by Sandell and Onishi [10], Iwantscheff [1], Schilt [28], Marczenko [36], Umland [37], Koch and Koch-Dedic [38], Lange and Vejdelek [39], Fries and Getrost [40].

The entries in Table 5 include mostly chelate complexes with large extinction coefficients. Azo dyes are representative of organic compounds which chelate with metal ions. Frequently, we can trace the effectiveness of dyes back to the formation of an ion association complex between a metal ion already complexed and one or several dye molecules. This complex can then be extracted from aqueous solutions with organic solvents and measured photometrically. Rhodamine B was the first basic dye to be used for this *extraction spectrophotometry*. For example, this dye forms an ion pair with

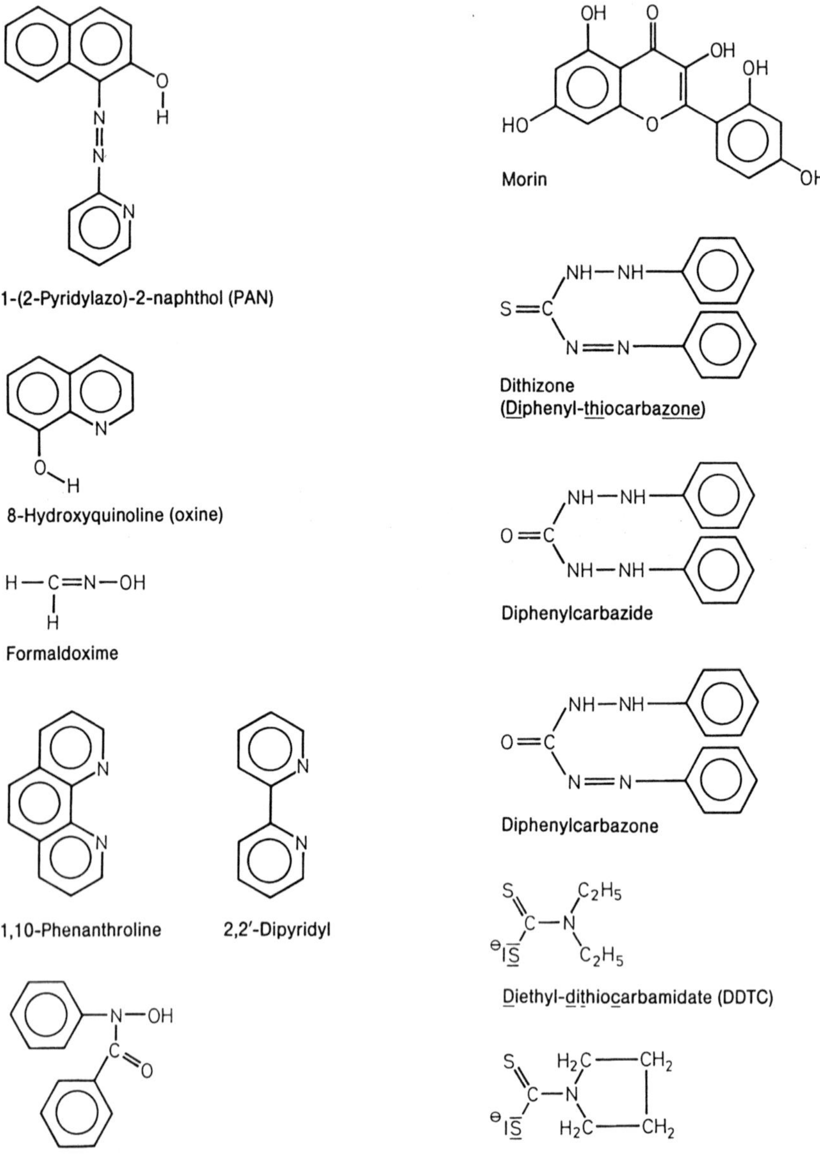

Fig. 9. Structural formulas of some complexing agents

$SbCl_6$ in acid solution (6N HCl) at a ratio of 1 : 1 and can be extracted with benzene [64–68].

Further examples are the ion associates of methyl violet with fluorotantalate [69], of dodecamolybdato phosphoric acid with safranin [70] or crystal violet [71] and of tetrafluoroborate with methylene blue [72] or

Table 5. Summary of photometric determinations of elements. The references refer to the bibliography of detailed analytical procedures

Element	Reagent	λ in nm	$\varepsilon\, 10^{-3}$ $l\, mol^{-1}\, cm^{-1}$	Ref.
Aluminium	Aluminon	525	11.0	36
	Eriochrome cyanine R	535	74.0	40
	8-hydroxyquinoline	386	6.6	36
	Salicylic acid	410		44
Antimony	Silver diethyl dithiocarbamidate	510	18.2	40
	Phenylfluorene	530	34.2	36
	Rhodamine B	565	85.0	40
	Pyrocatechol violet	555		45
	Bromopyrogallol red	560	37.4	36
Arsenic	Silver diethyl dithiocarbamidate	538	13.5	36, 40
	Determination as arsenic molybdenum blue complex			
Barium	Dimethylsulfonazo DAL	670	5.3	36
Beryllium	Acetylacetone	295	31.6	36
	Erichrome cyanine R	512	13.5	36
	Chromal blue G + Cetyltrimethylammonium chloride	626	93.0	46
Bismuth	Dithizone	495	80.0	36
	Sodium diethyl dithiocarbamidate	366	8.6	36
	Xylenol orange	550	11.0	40
Boron	Curcumin	555	146.0	36
	Carminic acid	610	5.7	40
	1,1'-dianthrimide	620	18.0	40
Cadmium	Dithizone	520	65.0, 85	36, 40
	Pyridylazonaphthol	550		36
Calcium	Glyoxal-bis (2-hydroxyanil)	520	16.31	36
	8-hydroxyquinoline/n-butylamine	370	6.37	36
Cerium	8-hydroxyquinoline	505	6.0	40
Chlorine	*o*-tolidine	440	28.0	40
Chromium	Diphenylcarbazide	540	34.0	40
Cobalt	1-nitroso naphthol-2	420	34.0	36, 40
	2-nitroso naphthol-1	530	14.7	36
	Nitroso-R salt	500	14.0	40
	1-(2-pyridylazo) resorcinol	510	56.7	48, 49
Copper	Bathocuproine	479	14.2	36
	Cuprizon	595	16.0	36, 40
	Sodium diethyl dithiocarbamidate	440	16.0	40
	Diphenylcarbazide	495	158.8	50, 51
Fluorine	Ce-chelate with alizarine-3-methyl-amine-*N,N*-di-acetic acid	617	13.7	36
	La-chelate ditto	620	11.0	40
Gallium	Xylenol orange	545	32.9	36
Germanium	Phenylfluorene	510	87.0	40
Gold	Rhodamine B	565	61.0	40
	Pyridine-2-aldoxime	420		36
Hafnium	Arsenazo (with Zr)	570		36
Indium	Dithizone	510	69.0	40
	4-(2-pyridylazo)-resorcinol	500	17.1	36
	Xylenol orange	560	25.9	36
Iodine	*o*-tolidine	425	20.0	40

Table 5 (continued)

Element	Reagent	λ in nm	ε 10^{-3} $l\,mol^{-1}\,cm^{-1}$	Ref.
Iridium	Pyridyl azonaphthol	550	10.3	36
Iron (II)	1,10-phenanthroline	512	11.1	36
	Bathophenanthroline	533	22.4	36, 40
	2,2′-dipyridyl	522	8.7	36
Iron (III)	Ferron (7-iodo-8-hydroxyquinoline-5-sulfonic acid)	610	5.7	36
Lanthanides	Alizarin S (total determination)	530–550		36
Lead	Dithizone	520	72.9	36, 40
	8-hydroxyquinoline in molten naphthalene /$CHCl_3$	360	91.0	47
Magnesium	8-hydroxyquinoline/n-butylamine	380	5.6	36
	Eriochrome black T	530	24.0	40
Manganese	Formamidoxime	450	11.0	40
	Diethyldithio carbamidate	500	4.0	36, 40
Mercury	Dithizone	485	68.0	36, 40
Molybdenum	6,7-dihydroxy-2,4-diphenyl benzopyrilium chloride	535	50.0	52
	toluene-3,4-dithiol	670	23.0	40
	Chloranilic acid	350	10.0	40
	Quercetin	420	36.0	40
	Thioglycolic acid	365	2.35	36
Nickel	Dimethylglyoxime	450	16.0	36
	Sodium diethyl dithiocarbamidate	325	35.0	40
	Pyridylazonaphthol	560	61.0	53–55
Niobium	4-(pyridyl-2-azo)-resorcinol	550	38.0	40
	Bromopyrogallol red	610	47.5	36
Osmium	1,5-diphenylcarbazide	560	140.0	40
	1-naphthylamine-4,6,8-trisulfonic acid	555	29.8	36
	Mercapto-benzimidazole	550	10.5	36
Palladium	2,2′-furildioxime	380	22.5	36
	2-nitroso-naphthol-(1)	370	22.3	36
Phosphorus	as phosphorus molybdenum blue	725	16.3	36
(PO^{3-})	as molybdato phosphoric acid	310	24.4	36
Platinum	Dithizone	720	38.0	36
Rhenium	2,2′-furildioxime			
	a) in aqueous phase	522	41.3	36, 40
	b) in organic phase	530	29.8	36
Rhodium	1-(pyridylazo)-naphthol-(2)	600	5.4	40
Ruthenium	Nitroso-R salt	580	22.2	36
	N,N′-diphenyl thiourea	650	0.23	40
Scandium	Alizarin S	520	5.4	36, 40
	8-hydroxyquinoline	378	6.9	36
Selenium	3,3′-diaminobenzidine	420	9.9	36
	o-phenylenediamine	355	17.8	36
	2,3-diaminonaphthalene	377	23.8	36
Silicon	as molybdato silicic acid	390	1.7	36
	as silico molybdenum blue	691 or 730		36
Silver	Dithizone	462	30.6	36, 40
Strontium	Arsenazo III	600		39

Table 5 (continued)

Element	Reagent	λ in nm	ε 10^{-3} l mol^{-1} cm^{-1}	Ref.
Tantalum	Pyrogallol	325	4.8	36
Tellurium	Sodium diethyl dithiocarbamidate	430	3.7	36
	Bismuth thiol	335	28.0	36
Thallium	Brilliant green, see Table 6	630	100.0	40
	Rhodamine	560	100.0	40
	Dithizone	505		36
Thorium	Thorin	545	10.2	36
	Arsenazo	660	130.0	40
Tin	Phenylfluorene	510	56.0	36, 40
	Pyridyl-3-fluorene	545	110.0	40
	Hematin	590	76.0	36
	Bromopyrogallol red	515	18.0	36
Titanium	Pyrocatechol disulfonic acid 3,5	410	13.6	40
	N-benzoyl-*N*-phenylhydroxylamine	380	6.7	36
Tungsten	Dithiol	640		36
Uranium	Glyoxal-bis-(2-hydroxyanil)	570	24.0	36, 40
	Pyridylazoresorcinol	530	38.7	36
	1-(pyridyl-2-azo)-naphthol-2	560	20.0	40
	2-(5-bromopyridyl-2-azo)-5-(diethylamino)-phenol	580	74.0	40
Vanadium	*N*-benzoyl-*N*-phenylhydroxylamine	510	4.7	36, 40
	Pyridylazoresorcinol	550	36.0	36
	Vanadox (2,2′-dicarboxydiphenylamine)	610	23.0	40
Yttrium	Alizarin S	550		56
	Pyrocatechol violet	665	25.9	57
	Dicarboxy arsenazo III	645	79.6	58
Zinc	Dithizone	538	95.0	36, 40
	Zincon[2-{|α(2-hydroxy-5-sulfophenylazo)-benzylidene|-hydrazino}-benzoic acid-monosodium salt]	625	24.0	40
Zirconium	Alizarin S	560		36
	Pyrocatechol violet	650	32.6	59, 61
	Xylenol orange	535	12.8	40, 61
		600	75.0	62
	Arsenazo III	665	120.0	63

monomethylthionine [73]. These ion associates with basic or acid dyes generally have very high extinction coefficients since several dye molecules can be present in the complex. On account of the demand for greater sensitivity and the resulting lowering of the detection limit into the ppb region, numerous investigations of extraction spectrophotometry have been carried out in recent years. Tables 6 and 7 provide a summary, compiled by Marczenko [74], of the elements extracted in the form of ion associates with basic or acid dyes.

Table 6. Summary of selected extraction spectrophotometric methods of determination with basic dyes [74]

Element	Complex	Dye	Solvent	ε_{max}	Specific absorption a
Antimony	$SbCl_6$	Rhodamine B	Benzene	97×10^3	0.80
	$SbCl_6$	Brilliant green	Toluene	103×10^3	0.85
	$SbCl_6$	Butylrhodamine B	Toluene/butanol	120×10^3	0.99
Bismuth	BiI_6	Rhodamine B	Benzene	130×10^3	0.62
Boron	BF_4	Methylene blue	Dichloroethane	65×10^3	6.01
	BF_4	Chromopyrazole II	Chloroform	67×10^3	6.20
Gallium	$GaCl_4$	Methylene blue	Benzene/$C_2H_2Cl_2$	75×10^3	1.07
	$GaCl_4$	Rhodamine B	*o*-dichlorobenzene	90×10^3	1.29
Germanium	Dinitro pyrocatechol	Brilliant green	CCl_4	141×10^3	1.94
	Alizarin complexone	Rhodamine 6 G	$CCl_4/CHCl_3$	290×10^3	4.00
Gold	$AuCl_4$	Rhodamine B	Benzene	97×10^3	0.49
	$AuCl_4$	Methyl violet	Trichloroethylene	115×10^3	0.58
Indium	InI_4	Malachite green	Benzene/hexane	106×10^3	0.92
	$InBr_4$	Rhodamine B	Benzene/diiso propylether	110×10^3	0.96
Mercury	$HgCl_4$	Crystal violet	Toluene	55×10^3	0.27
	$HgCl_4$	Methyl green	Benzene/toluene	131×10^3	0.65
Phosphorus	Hetero-polys.	Malachite green	Propylacetate	170×10^3	5.49
	Hetero-polys.	Crystal violet	Propylacetate	270×10^3	8.72
Rhenium	ReO_4	Butylrhodamine B	Benzene	40×10^3	0.21
	ReO_4	Nile blue	Chlorobenzene + methanol		
Tantalum	TaF_6	Nitrochromopyrazole	Benzene/toluene	83×10^3	0.46
	TaF_6	Methylene blue	Dichloroethane + trichloroethylene	91×10^3	0.50
	TaF_6	Capri blue	Chloroform	107×10^3	0.59
Tellurium	$TeBr_6$	Victoria blue 4R	Benzene + nitro-benzene	83×10^3	0.63
Thallium	$TlCl_4$	Crystal violet	Diisopropylether	102×10^3	0.50
	$TlCl_4$	Brilliant green	Diisopropylether	106×10^3	0.52
	$TlCl_4$	Methylene blue	Dichloroethane + trichloroethane	114×10^3	0.56
Tin	$SnCl_6$	Crystal violet	Heptanone	85×10^3	0.72
Tungsten	Dinitro pyrocatechol	Brilliant green	Chloroform	132×10^3	0.72
			Benzene + MIBK	103×10^3	0.43
Uranium	Benzoic acid	Rhodamine B			

The solvents listed in Tables 6 and 7 refer to the solvent used for extracting the ion associate. If the extinction coefficients of the ion associates in these tables are compared with the corresponding entries in Table 5 it can be seen that, in some cases, very high extinction coefficients are actually observed which implies a high sensitivity of detection. For the evaluation of

Table 7. Extraction spectrophotometric methods of determination with acid dyes; metals complexed with 1,10-phenanthroline [74]

Dye	Element	Solvent	ε_{max}	Specific absorption
Rose Bengal B	Copper (II)	Chloroform	63×10^3	0.99
		Ethyl acetate	78×10^3	1.23
	Palladium	Chloroform	50×10^3	0.47
	Zinc	Chloroform	51×10^3	0.78
	Cadmium	Chloroform	64×10^3	0.57
	Lead	Chloroform	58×10^3	0.28
Eosin	Silver	Nitrobenzene +	55×10^3	0.51
		chloroform +		
		butanol	50×10^3	0.46
	Lanthanides	Toluene + butanol	120×10^3	
	Zinc	Chloroform	120×10^3	1.84
	Lead	Chloroform	110×10^3	0.53
Erythrosin	Lanthanides	Toluene + butanol	160×10^3	
	Lead	Chloroform	65×10^3	0.31
Bromphenol blue	Iron (II)	Chloroform + amylol	59×10^3	1.06
		Nitromethane	82×10^3	1.47
	Silver	Chloroform		
	Zinc	Chloroform	100×10^3	1.53
	Zinc	Chloroform	38×10^3	0.58
	Cadmium	Dichloroethane	31×10^3	0.28
Bromphenol red	Iron (II)	Nitromethane	56×10^3	0.57
	Zinc	Chloroform + amylol	32×10^3	0.49
Bromcresol green	Zinc	Chloroform	17×10^3	0.29
Methyl orange	Iron (II)	Chloroform	48×10^3	0.86

the sensitivity, the values of the specific absorption "a" are also given for each element calculated from Eq. (16).

It has been found that some low-solubility ion associates of basic dyes with anion complexes cannot be extracted by means of organic solvents which have low dielectric constants, because they coagulate into flakes which collect at the phase boundary or adhere to the walls. The solvent and aqueous solution are separated by decanting or filtering and the precipitate is then dissolved (usually in acetone or alcohol). The complex then dissociates and the procedure forms the basis of a new photometric technique which is called *flotation spectrophotometry.*

The high extinction coefficients determined with this method depend on the dissociation of the low-solubility ion associate which dissociates upon dissolution. Subsequently, the absorbance of the *dye* transferred into solution is measured in the solution. Since several dye molecules can be fixed in a complex and since the measured absorbance is related to the molecular weight of the complex, extremely high extinction coefficients are obtained per gram atom of the element to be determined. Whole-number multiples of the extinction coefficient of the dye in the same solvent are involved. For example, rhodium as $RhCl_3$ with $SnCl_2$ results in an ion associate in which

Table 8. Flotation spectrophotometric methods for determining elements from Marczenko [74]

Element	Complex with	Basic dye	Solvent flotation	Solution	ε $1\,mol^{-1}\,cm^{-1}$	
Arsenic	Mo-As-O	Crystal violet	Cyclohexane + toluene	Acetone	320	10^3
Bismuth	Br^-	Rhodamine 6G	Diisopropyl-ether	Ethanol	150	10^3
Cadmium	I^-	Crystal violet	Diisopropyl-ether	Acetone	130	10^3
Germanium	Mo-Ge-O	Brilliant green	Butylacetate	Acetone	190	10^3
		Rhodamine 6G	Toluene	Ethanol	310	10^3
		Rhodamine B	Toluene	Ethanol	370	10^3
	Alizarin complexone	Rhodamine 6G	$CCl_4 + CHCl_3$	Ethanol	290	10^3
Silicon	Mo-Si-O	Crystal violet	Propylacetone	Acetone	140	10^3
		Rhodamine B	Diisopropyl-ether	Ethanol	230	10^3
Molybdenum	SCN^-	Crystal violet	Toluene	Ethanol	230	10^3
Phosphorus	Mo-P-O	Crystal violet	Butylacetate	Methyl-ethylketone		
Rhodium	$RhCl_3SnCl_2$	Malachite green	Diisopropyl-ether	Acetone	340	10^3
Tellurium	Br^-	Rhodamine 6G	Benzene	Ethanol	170	10^3
Zirconium	Picramin-epsilon	Ethylrhodamine B	Benzene	Acetone	320	10^3

five malachite green molecules are bound. To this fact the extinction coefficient $\varepsilon = 340 \times 10^3\,1\,mol^{-1}\,cm^{-1}$ quoted for this method of determination may be ascribed [75]. In the determination of osmium, an ion associate of $Os(SCN)_6^{3-}$ and *three* methylene blue molecules is formed [76]. The extinction coefficient $\varepsilon = 220 \times 10^3\,1\,mol^{-1}\,cm^{-1}$ at 655 nm is dependent on the *three* dye molecules. $Pd(SCN)_4^{2-}$ forms an ion associate with methylene blue (MB) which is formulated of $[MB^+]_2[Pd(SCN)_4^{2-}]$. Thus, the measured extinction coefficient corresponds to *two* methylene blue molecules [77].

A synopsis of *flotation spectrophotometry* is given in Table 8. The composition of the ion pairs depends very strongly on the experimental conditions and particularly on the solvents used for flotation and dissolution [74].

The methods described operate discontinuously but a continuous method for the determination of traces of chromium(VI) has also been reported [78]: sodium lauryl sulfate forms an ion pair with a complex of chromium diphenyl carbazone which concentrates at the boundary surface of the foam produced in this method. This ion pair can be measured spectrophotometrically either discontinuously or continuously after diluting the foam. In many cases, Cr(VI) can be measured with an accuracy of $\pm 3\%$ even if there is a hundredfold excess of foreign ions.

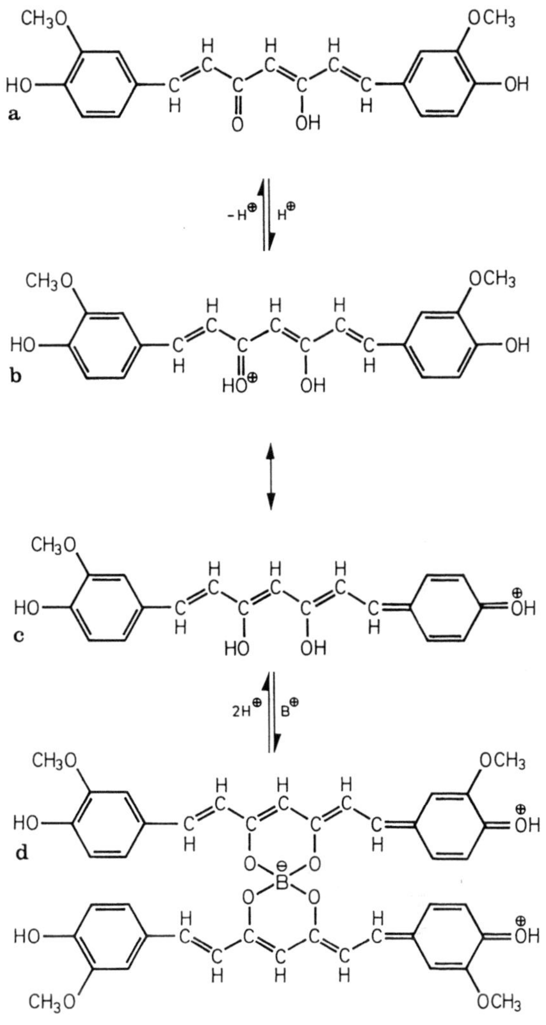

Fig. 10 a–d. Complexing mechanism for boron with curcumin (see text for explanation)

Another application of the method depends on the water-soluble ion association complexes which metal ions and xylenol orange (R) form with cetylpyridinium bromide (CP). For example, the complex $La[R(CP)_2]_2$ which has an extinction coefficient of $\varepsilon = 92 \times 10^3\ l\ mol^{-1}\ cm^{-1}$ at 625 nm is formed with lanthanum [79].

The development of new methods generally has the goal of converting systems with low extinction coefficients into those with the highest possible ones, even though the extremely high extinction coefficients are only consequences of the procedure in flotation spectrophotometry.

Another particularly interesting method is based on the intensive color occurring in chelates due to the stabilization of a *particular ligand structure* where the complexed element has no influence upon the position and intensity of the spectrum. The photometric determination of boron with curcumin is an example [80]. Curcumin is the enol of a 1,3-diketone which changes into a mesomeric form in a strongly acid solution. This form shows a continuous conjugation with charge resonance and has long wavelength bands at $18020\,cm^{-1}$ with extinction coefficient $\varepsilon_{max} = 73.6 \times 10^3\,l\,mol^{-1}\,cm^{-1}$. Mesomeric forms are shown in Fig. 10.

The mesomeric extreme form c has a structure with two enolic OH-groups which is selective for boron. This structure is fixed by ester chelate formation if boric acids are present. The position of the absorption maximum and extinction coefficient of this 1 : 1 complex is similar to that of the protonated molecule [81]. However, since boron is capable of binding two curcumin ligands the extinction coefficient of this 1 : 2 complex is then twice as large. Instead of $\varepsilon = 73.6 \times 10^3\,l\,mol^{-1}\,cm^{-1}$ for the chelate with *one* curcumin ligand in ethanol as solvent we obtain $\varepsilon = 146 \times 10^3\,l\,mol^{-1}\,cm^{-1}$ for the chelate with *two* curcumin ligands.

Such a large extinction coefficient ensures a sensitivity in the nanogram region for the determination of boron; and in conjunction with paper chromatography, this method can be extended quantitatively into the picogram region [82, 83].

The operating conditions must be strictly observed when using this technique. This applies particularly to the blank value in the photometric measurements. For this reason, Tôei et al. proposed another method for determining boron [84] which is based on the extraction spectrophotometry described above. In this case, boron is complexed with 2,4-dinitronaphthalene-1,8-diol and converted into an ion associate with brilliant green which is extracted into toluene and measured photometrically. The complex has an extinction coefficient of $\varepsilon = 103 \times 10^3\,l\,mol^{-1}\,cm^{-1}$ at $\lambda = 693$ nm, i.e. this method is also very sensitive and, furthermore, very selective for boron [84].

.2 Photometric Determination of Anions and Ammonia

Every experimental technique aims at a conversion into a colored species. Table 9 shows that we use reactions which primarily involve the anion to be determined. However, it need *not* necessarily be present in the colored end product. The determination of bromide and iodide are exceptional because they can be determined directly in organic solvents after oxidation to bromine or iodine. The oxidation to the iodate ion also plays a role in the case of iodide where we use a reaction with an added ethanolic KI solution, i.e. the formation of I_2, for a photometric determination. If bromine is allowed to react with fluorescein, tetrabromofluorescein (i.e. eosin) is formed which we can easily measure photometrically; and similarly bromophenol blue is formed from phenol red and bromine. Another example of the *inclusion of the anion to be determined in an organic compound* with a change in the absorption spectrum of the organic component is nitrate

Table 9. Classification of the photometric determination of anions, from Lange and Vejdelek [39]

Anion	Reagents; reaction	Determination as	nm	
Br^-	1) K_2CrO_4 KMnO4:	oxidation to Br_2	Br_2	400 – 410
	2) Chloramine T	Br_2 + fluorescein	Eosin	
	3) Chloramine T	Br_2 + phenol red	Bromphenol blue	580 – 610
	4) $AuCl_3$		K $AuCl_3Br$	350; 440 – 470
Cl^-	1) Mercuric chloranilate		Chloranilate ion	305; 530 – 540
	2) $AgNO_3$	AgCl; Ag^+ + dithizone	Ag-dithizonate	595 – 600
	3) Ag_2CrO_4	AgCl; CrO_4^{2-} + 1,5-diphenylcarbazide	Cr(III)-diphenylcarbazone	540
	4) Ag_2CrO_4	AgCl; CrO_4^{2-} + H_2O_2	CrO_7^{4-}	620
	5) $Hg(NO_3)$	Hg; Hg + 1,5-diphenylcarbazide	Complex	520
	6) $Hg(SCN)_2$	Hg_2Cl_2 + SCN^-; SCN^- + Fe(III)	$Fe(SCN)_3$	460; 480 – 490
	7) Oxidation	Cl_2 + *o*-toluidine	Quinone derivative	435
CN^-	1) Phenolphthalein + $CuSO_4$ (pH = 11) as catalyst		Phenolphthalein	530 – 540
	2) Chloramine T	ClCN; ClCN + barbituric acid pH = 7	Polymethine dye	575 – 580
F^-	1) Zirconium alizarin complex pH 0.7 – 2.7		Decoloration	520 – 530
	2) Thorium (IV) alizarin complex pH 0.7 – 2.7		Decoloration	520
	3) La complex with alizarin complexone; double complex		La chelate	600 – 625
	4) CeIII complex with alizarin complexone; double complex		Ce chelate	600 – 625
	5) $Th(IO_3)_4$ IO_3^-; IO_3^- + iodide starch reagent		Starch-iodine complex	620 – 630
I^-	1) Oxidation I_2;	extraction	I_2 (oxidation solvent)	350 – 420
	2) Ce(IV)/As(III);	redox reaction	Ce(IV) decoloration	400 – 420
	3) H_2O_2 I_2;	I_2 + *o*-toluidine	Color reaction	pH 1.5 – 2; 420 pH 3.5 – 5; 620
	4) $KMnO_4$ (pH 3.5 – 4) IO_3;	IO^- + KI^- + 6 H^+ 3 I_2 + 3 H_2O	$I_2(CCl_4, CS_2$, etc.)	310; 350 – 360
	4 a)	+ KI (ethanol)	KI_3 solution	350 – 360
	4 b)	+ starch solution	Starch-iodine complex	570

Table 9 (continued)

Anion	Reagents; reaction	Determination as	nm
NO_3^-	1) $FeSO_4$ in conc. H_2SO_4;	$(FeNO)SO_4$	520 – 530
	2) Xylenols in $H_2O + CH_3COOH + H_2SO_4$ nitroxylenols	Nitroxylenol	410 – 430
	nitroxylenols	Nitrosoxylenol	410 – 430
	3) Brucine in $H_2O + H_2SO_4$ color reaction	Reaction product	400 – 420
	4) Phenol disulfonic acid; 5-nitrophenol-2,4-disulfonic acid	Reaction product	400 – 425
	5) Chromotropic acid in conc. H_2SO_4	Reaction product	357; 400 – 430
	6) Sodium salicylate + $H_2O + H_2SO_4$	5-nitrosalicylic acid	410 – 420
NO_2^-	1) Sulfanilamide; diazotization; coupling to azo dye with *N*-(1-naphthyl)-ethylenediamine 2HCl	Azo dye	540 – 545
	2) Naphthylamine diazotization; + naphthylamine	Azo dye	560
	3) Thio Michler ketone	Color reaction	650
Phosphates	see Table 6 under element phosphorus		
Silicates	see Table 6 under element silicon		
SO_4^{2-}	1) $BaCrO_4$ $BaSO_4 + CrO_4^{2-}$; CrO_4^{2-} + diphenylcarbazide	Cr(III)-diphenylcarbazone	540
	2) Benzidine benzidine sulphate		
	a) $H_2O_2 + FeCl_3$	Reaction product	410 – 430
	b) Diazotization + coupling with thymol	Azo dye	
	c) Exchange with sodium 2-naphthoquinone-4-sulfonate	Reaction product	470 – 490
	3) Barium chloranilate $BaSO_4$ + chloranilate ion	Chloranilate ion	305; 530 – 540
	4) 2-amino-pyrimidinium chloride 2-amino-pyrimidinium sulfate	Reagent excess	305; 525
	5) Reduction to S^{2-}; + *N,N*-dimethyl-*p*-phenylenediamine, Fe(III) salts as catalyst	Methylene blue	600 – 670
H_2S; S^{2-}; S	see SO_4^{2-} method 5)		
$S_2O_3^{2-}$,	Alkali cyanide in the presence of Cu(II) salts		
$S_3O_6^{2-}$, $S_4O_6^{2-}$	SCN^-; SCN^- + Fe(III)	$Fe(SCN)_3$	460

detection with phenols where yellow nitrophenol derivatives are formed in the presence of concentrated sulfuric acid [85–87]. By controlling the reaction conditions, the nitrate ion can be determined photometrically as 5-nitrosalicylic acid with sodium salicylate. In the case of the nitrite ion, the diazotizing reaction with subsequent coupling to an azo dye is used for photometric determination. This reaction can also be used for nitrate determination if we first reduce the nitrate ion to nitrite [88, 89].

Another reaction where the anion to be determined is incorporated into the end product is the detection of S^{2-} or H_2S. Here, methylene blue, which can conveniently be measured photometrically, is formed from *N,N*-dimethyl-*p*-phenylendiamine in the presence of Fe(III) salts as catalyst. If SO_4^{2-} ions are initially reduced to S^{2-} ions, this method can also be used for sulfate determination [90–93].

All the other methods shown in Table 9 such as those for chloride, fluoride, cyanide and sulfate ions, are indirect techniques which, in some cases, are based on the precipitation of difficultly soluble compounds of these anions. Equivalent quantities of a second ion, or even of a cation, are released which can be measured photometrically, in some cases with great sensitivity, by means of characteristic color reactions.

Lange and Vejdelek have collected the photometric methods of analysis in their monograph [39].

All these methods suffer from the defect that the initial chemical reactions are very time-consuming.

A rapid determination of *nitrate ion* has been described which leads to a reduction product with 4,5-dihydroxycoumarin via the reduction of the nitrate ion to nitrite. This product can be measured photometrically at 410 nm [94, 95]. Baca and Freiser [96] proposed an extraction-photometric method for determining nitrate ion which is based on the formation of an ion pair between the nitrate ion and crystal violet; sensitivity to 0.06–0.72 ppm nitrate is reported. Flamers and Bashier [97] modified the azo dye method for the microdetermination of nitrate ion; 0.003 $\mu g/cm^3$ can be detected.

There are several techniques for determining *NH_3 and ammonium salts* [39]. The indophenol reaction which is specified for the measurement of NH_4^+ or NH_3 in drinking, industrial and waste water is the best known of these [98, 99].

The *o*-tolidine method is very important for the determination of *halogens in water*. In this analysis, free chlorine or bromine in aqueous solution oxidises *o*-tolidine to produce a quinonoid system which can be determined photometrically. The quinonoid compound has an absorption maximum at $\lambda = 435$ nm. Analogous reactions with diethyl-*p*-phenylenediamine and syringaldazine are further methods. In the case of syringaldazine, a hydroquinone derivative, we obtain a red quinone as an oxidation product with an absorption maximum at $\lambda = 530$ nm. Corresponding reactions are shown in reaction scheme A, a–c.

Reaction scheme A, a–c

a) *o*-tolidine: (X = Cl, Br)

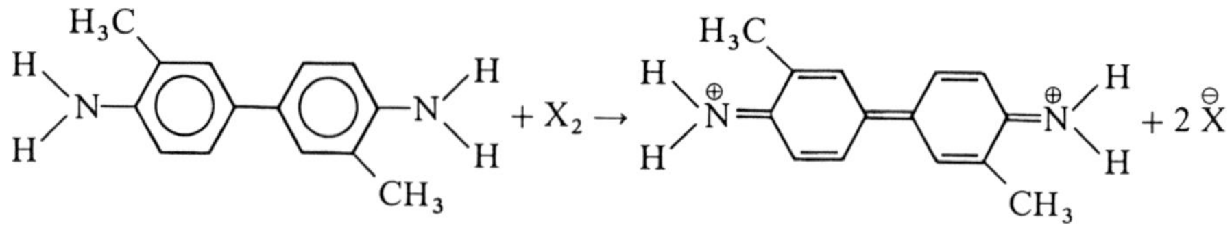

b) Diethyl-*p*-phenylenediamine:

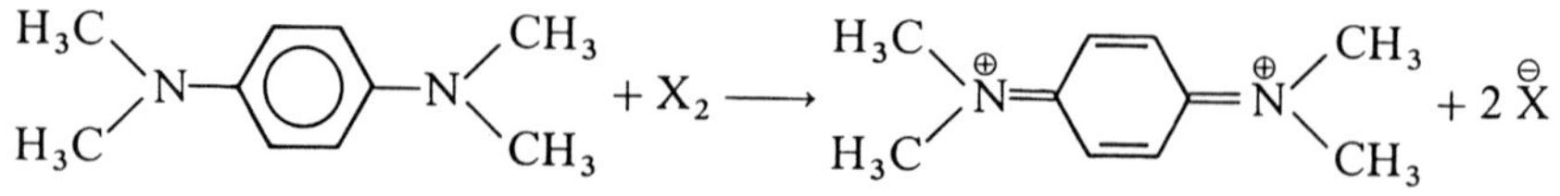

c) Syringaldazine:

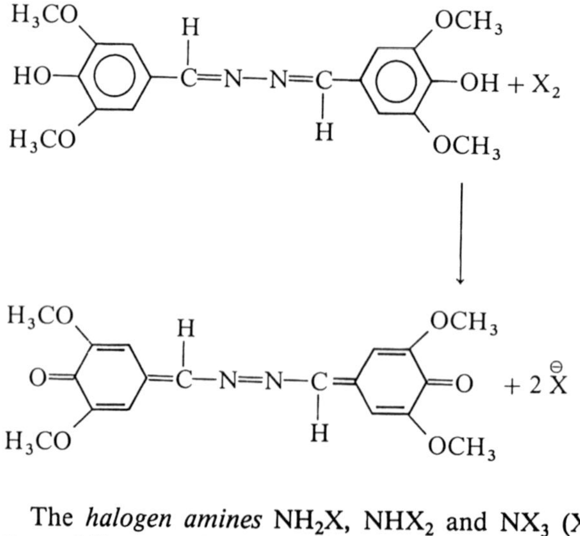

The *halogen amines* NH_2X, NHX_2 and NX_3 (X = Cl, Br), which are formed if ammonia and halogens are present in the water to be analyzed, also undergo the same reactions. Soulard et al. [100] investigated all three methods critically. They concluded that the *o*-tolidine method allows a simple, sensitive and accurate determination of the total halogen content in a solution, but it cannot differentiate between a free halogen and a halogen bound in the form of halogen amines. Provided that specific experimental conditions are observed, the diethyl-*p*-phenylenediamine method permits differentiation. However, it is less sensitive on the whole. Results for the syringaldazine method are similar to those obtained with *o*-tolidine.

4.1.3 Photometric Water Analyses

Methods for determining cations and anions in water were laid down in the German drinking water regulations of 31. 10. 1975 [98]. The Standardization Committee responsible for water quality recently prepared "Guidelines for setting up sampling programmes" [99], an extract from which is shown in Table 10. The applicable concentration range in ppm is given in the second column. The monograph of Freier should also be consulted [102]. Hein has reviewed water analyses by UV-VIS spectrophotometry [103].

Franke and Hein have given a general review of water analysis with spectroscopic and chromatographic methods [104].

On account of the general importance of water analyses, the filter photometer 'Nanocolor 25' should be mentioned because it is a simple instrument for routine analyses of surface and waste water. Test-analysis sets and operating instructions are available for the practical application of this

Table 10. Recommended working methods for the photometric determination of anions and cations in water

	Concentration range in ppm	Reagent	λ/nm
B	0.01 – 1	Azomethine	414
Cl^- (Cl_2)	0.05 – 25	*N,N*-diethyl-*p*-phenylenediamine	510 + 550
CN^-	0.002 – 0.02	Barbituric acid-pyridine	578
F^-	0.02 – 2	Lanthanum alizarin complexone	610
I^-	0.001 – 0.007	Redox system CeIV/As(III)	
SiO_2	0.1 – 10	As silicomolybdic acid	720
NO_3^-	0.1 – 10	Sodium salicylate	420
NO_2^-	0.001 – 0.3	Sulfonylamide + *N*-(1-naphthyl-ethylenediamine)	530
PO_4^{3-}	0.002 – 0.6	As phosphorus molybdenum blue	750
SCN^-	0.05 – 50	Pyridine benzidine	491
SO_4^{2-}	2 – 60	As $BaSO_4$ in gelatine solution (measurement of light scattering)	490
S^{2-}	0.01 – 5	Dimethyl-*p*-phenylenediamine	670
Al	0.02 – 0.7	Eriochrome cyanine R	530
As	0.002 – 0.1	Silver diethyldithiocarbamidate	546
NH_4^+	0.005 – 2	Indophenol	690
Pb	0.002 – 20	Dithizone	520
Cd	0.002 – 20	Dithizone	530
Cr	0.005 – 10	Diphenylcarbazide	550
Fe	0.01 – 4	1,10-phenanthroline	510
Cu	0.001 – 0.3	Zn-*N,N*-dibenzyldithiocarbamidate	436
Mn	0.01 – 5	Formaldoxime	480
Ni	0.02 – 10	Diacetyldioxime	
Se	0.001 – 0.25	*o*-phenylenediamine	334
Ag	0.05 – 2	Dithizone	470
U	0.001 – 0.01	Arsenazo-III	665
V	0.05 – 40	*N*-benzoyl-*N*-phenylhydroxylamine	546
Zn	0.004 – 20	Dithizone	530

photometer [105]. Similar test-analysis sets, not only for water analyses, are supplied for their filter photometers by the Dr. Lange Company.

In practical water analysis, it is especially important to monitor the silicic acid concentration in boiler feed water in order to ensure the operational safety of steam generating plants. The Polymetron Company have developed the photometer type 8570 for this purpose. It is a module of the "Silkostat" which allows continuous monitoring of silicic acid in clean water. The determination is made with silicomolybdenum blue (see Table 5) and takes about 6–7 min.

4.1.4 Photometric Determination of Organic Compounds

Organic compounds with a chromaphoric system absorb in the UV-VIS region. Since it is usually assumed that their spectroscopic data are known, the determination of single substances is always possible by means of the Bouguer-Lambert-Beer law. However, attention must be paid to the solvent because the position and intensity of the absorption maxima may depend to a great extent upon the solvent. Furthermore, the influence of the pH value of the solution must be taken into account for basic and acidic compounds.

The quantitative determination of a single organic compound is often made difficult by the fact that other compounds may be present in the system to be analyzed. The absorption spectra of these other compounds overlie that of the substance under investigation. An attempt may be made to separate the mixture. However, if the absorption spectrum of each individual compound is known then the composition of the mixture can be determined accurately by means of a multicomponent analysis (cf. Sect. 4.2).

One of the organic components of a mixture often has a relatively low extinction coefficient which does not allow accurate determination. This applies, for example, to saturated ketones, aldehydes, carboxylic acids and their derivatives ($\varepsilon < 50\,\mathrm{l\,mol^{-1}\,cm^{-1}}$) which also absorb in the analytically unfavorable region below 300 nm. In this event it may be possible to produce a derivative whose absorption spectrum is shifted bathochromically and has a greater intensity and thus differs from that of the other components. The aldehyde determination with derivatives of *phenylhydrazine* is an example. Here a phenylhydrazone, whose spectrum shows relatively large extinction coefficients, is easily formed [106, 107];

$$O_2N\text{-}C_6H_3(NO_2)\text{-}NH\text{-}NH_2 + O{=}C\begin{matrix}R_1\\R_2\end{matrix} \longrightarrow O_2N\text{-}C_6H_3(NO_2)\text{-}NH\text{-}N{=}C\begin{matrix}R_1\\R_2\end{matrix}$$

Table 11. Absorption maxima of 2,4-dinitro phenylhydrazones of aldehydes and ketones in alcohol [111]

$X = O_2N-C_6H_3(NO_2)-$, R = Alkyl

Substance	λ_{max} A nm	$\tilde{\nu}_{max}$ cm^{-1}	$\varepsilon \cdot 10^{-3}$
$X-NH-N=H_2$	350	28550	15
$X-NH-N=CH_2$	348	28750	18.2
$X-NH-N=CHR$	356–360	27800–28100	20–30
$X-NH-N=CRR'$	360–365	27400–27800	
$X-NH-N=C<(CH_2-CH_2)(CH_2-CH_2)$	363	27550	21.5
$X-NH-N=CH-CH=CH_2$	366	27300	25–35
$X-NH-N=CH-CH=CHR$	373–377	26550–26800	
$X-NH-N=CR-CH=CHR'$	376	26600	
$X-NH-N=CH-CR=CHR'$			
$X-NH-N=CH-CH=CRR'$	377–385	26000–26550	
$X-NH-N=CR-CH=CR'R''$	377–379	26400–26550	
$X-NH-N=CH-CR=CR'R''$	387	25850	
$X-NH-N=CH-CH=CH-CH=CHR$	379–395	25300–26400	30–40
$X-NH-N=CR-CH=CH-CH=CHR'$			
$X-NH-N=C<(CH=CHR)(CH=CHR')$			
$X-NH-N=CH-(CH=CH)_2-CH=CHR$	395–410	24400–25300	40–50
$X-NH-N=CR-(CH=CH)_2-CH=CHR'$			
$X-NH-N=CH-C_6H_5$	378	26450	29.2
$X-NH-N=C(C_6H_5)-C_6H_5$	383	26100	28.3
$X-NH-N=CR-C_6H_4-R$	383	26100	27.6
$X-NH-N=CH-C_6H_4(OH)$ (o-HO)	387	25850	29.5
$X-NH-N=CH-CH=CH-C_6H_5$	394	25400	38
$X-NH-N=CH-C_6H_4-OH$	395	25300	28.7
$X-NH-N=C(C_6H_5)-CH=CH-C_6H_5$	395	25300	36.4

Table 11 shows the absorption maxima of some phenyl-hydrazones of aldehydes and ketones and their associated extinction coefficients [108–110].

Figure 11 illustrates the absorption spectra of the parent phenylhydrazine as well as of some phenylhydrazones. Here dramatic changes of the absorp-

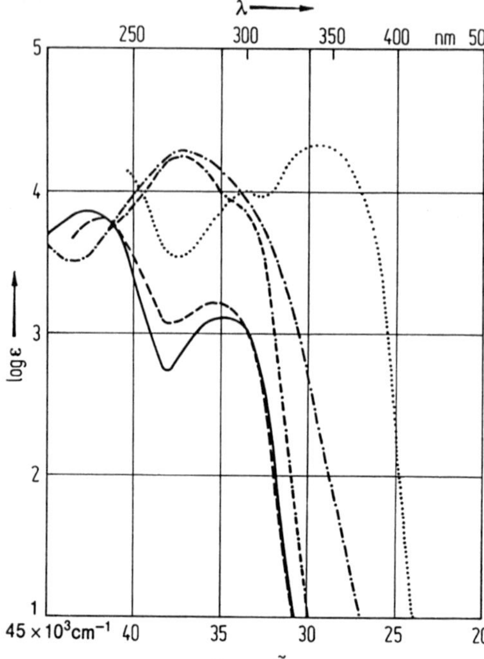

Fig. 11. Absorption spectra of selected phenylhydrazones. From Ref. [111]. Aniline in methanol (———), phenylhydrazine in ethyl alcohol (– – –), phenylhydrazone of acetaldehyde (–.–.–), benzaldehyde-phenylhydrazone (. . .), trimethylacetophenone-phenylhydrazone (– –.– –)

tion properties are seen [111]. The example of trimethylacetophenone phenyl-hydrazone (Fig. 11) also shows the steric influence of the *t*-butyl group which causes a strong hypsochromic shift of the absorption spectrum as can be seen when comparing this with the absorption spectrum of benzaldehyde phenylhydrazone (Fig. 11).

Phenylsemicarbazide or *2,4-dinitrophenylsemicarbazide* are further compounds which produce an extension of the conjugated system when coupled with aldehydes or ketones and thus effect a strong intensity increase. In this case, the corresponding phenylsemicarbazones are formed:

$$O_2N-C_6H_3(NO_2)-NH-\underset{\underset{O}{\|}}{C}-NH-N=C\begin{matrix}R_1\\R_2\end{matrix}$$

The corresponding semicarbazones are formed with unsubstituted semicarbazide (cf. Fig. 12) [111].

A sensitive detection of *acetaldehyde* can be achieved with 3-methyl-2-benzothiazolinone-hydrazone (MBTH). A polymethine dye is obtained which has a extinction coefficient of $\varepsilon = 76 \times 10^3\,l\,mol^{-1}\,cm^{-1}$ in acetone at $\lambda_{max} = 670\,nm$; see the diagram on p. 47, 48 [112, 113]. This reaction is also important for the photometric determination of *olefins*. Initially, an

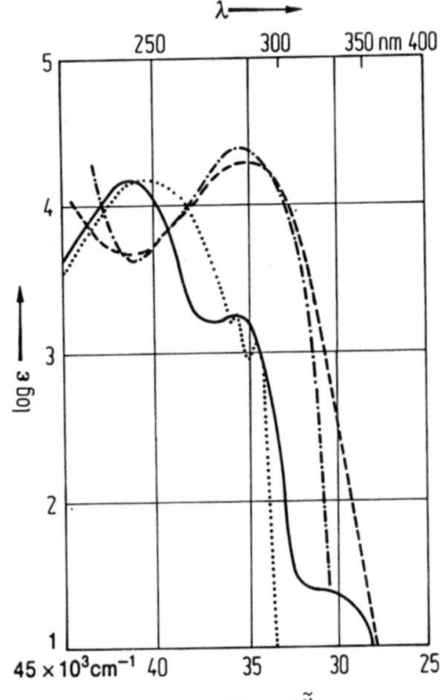

Fig. 12. Absorption spectra of semicarbazones. From Ref. [111]. Benzaldehyde (———), benzaldehydroxime (. . .), benzaldehydesemicarbazone (–.–), benzaldehyde-*N*-phenylsemicarbazone (– – –)

olefin of type $R-CH=CH_2$ is oxidized to an aldehyde and subsequently reacted with MBTH [113–116].

There are also numerous organic compounds from which *formaldehyde* can be obtained as a degradation product. This, when reacted with hydrazones, permits an 'indirect' determination. The monograph by E. and C. R. Sawicki contains a complete review of the photometric determination of aldehydes including compounds which can be regarded as aldehyde precursors [117].

These brief observations should show that the principle of spectrophotometric determination of organic compounds – as with inorganic compounds – lies essentially in producing, from parent compounds, substances whose absorption spectra have bands with large extinction coefficients shifted bathochromically.

1. (3-methylbenzothiazolyl, N-CH_3, S) $C=N-NH_2$ + CH_2O $\longrightarrow$ (3-methylbenzothiazolyl, N-CH_3, S) $C=N-N=CH_2$ P_1

2. (3-methylbenzothiazolyl, N-CH_3, S) $C=N-NH_2$ $\xrightarrow{O_2}$ (3-methylbenzothiazolyl, N-CH_3, S) $C=N-\overset{\oplus}{N}H$ P_2

3. $P_1 + P_2 \longrightarrow$

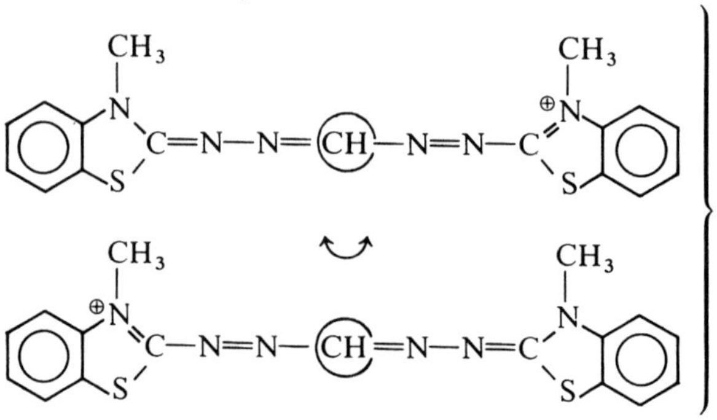

CH_3 / N / C=N—N=(CH)—N=N—C / $\oplus$N / CH_3 / S ⟷ CH_3 / $\oplus$N / C—N=N—(CH)=N—N=C / N / CH_3 / S

$\lambda_{max} = 670$ nm

$\varepsilon_{max} = 76 \cdot 10^3$
(Acetone)

Since it is not possible to discuss in detail here the photometric analysis of the following groups of organic compounds, the reader is referred to the monograph of Kakac and Vejdelek [118] for further information on:

Unsaturated hydrocarbons,
Hydroxy compounds,
Thiols and structurally related compounds,
Oxo compounds,
Carboxylic acids and their derivatives,
Organic sulfates and sulfonates,
Amino compounds,
Hydroxylamines,
Hydrazines,
Azo and diazo compounds,
Nitro and nitroso compounds,
Halogen compounds,
Organic metal and non-metal compounds,
Nitrogen-free heterocycles
Saccharides and their compounds,
Amino acids, peptides and proteins,
Steroids and structurally related compounds.

In recent decades, the photometric determination of organic compounds has gained particular importance in pharmaceutical, clinical, bio- and food chemistry and in problems of environmental protection.

The laboratories of the chemical industry and the equipment manufacturers have contributed significantly to progress in this field by continually developing test-analysis sets with exact operating instructions, see [119–121].

Lange and Vejdelek have compiled instructions for the photometric determination of organic compounds [39]. For the determination of organic compounds in water see [122, 129]. Pesez and Bartos have given further instructions [124]. Monographs are available for the analysis of vitamins [125–129] and steroids [130].

4.1.5 Enzymatic Analysis and Enzyme Kinetics

Enzymatic analysis, as a modern and versatile technique, has become increasingly more significant in food, clinical-chemical and biochemical analysis. Spectrophotometric monitoring of the kinetics of enzyme-controlled reactions is usually involved. This technique has been extended by the fitting of enzyme-kinetics accessories to spectrophotometers and photometers. One of the basic reactions can be described as follows:

$$\text{glucose-6-phosphate} + \text{NAD} \xrightleftharpoons{\text{G-6-PDH}} \text{6-phosphogluconolactone} + \text{NADH}$$

In this case, glucose-6-phosphate is oxidized to 6-phosphogluconolactone whilst nicotinamide-adenine-dinucleotide-diphosphate (NAD or NADP) is reduced to dihydronicotinamide-adenine-di-nucleotide-diphosphate (NADH or NADPH). The speed of formation of coenzyme NADH is proportional to the concentration of the catalyst, in this example the enzyme of glucose-6-phosphate-dehydrogenase (G-6-PDH). Since NADH has an absorption maximum at $\lambda_{max} = 340$ nm, the course of a reaction can be measured spectrophotometrically [119, 131], i.e. the absorbance measured at 340 nm increases as the above reaction proceeds from left to right.

We can follow hydrogen addition to NAD or abstraction from NADH analytically as follows:

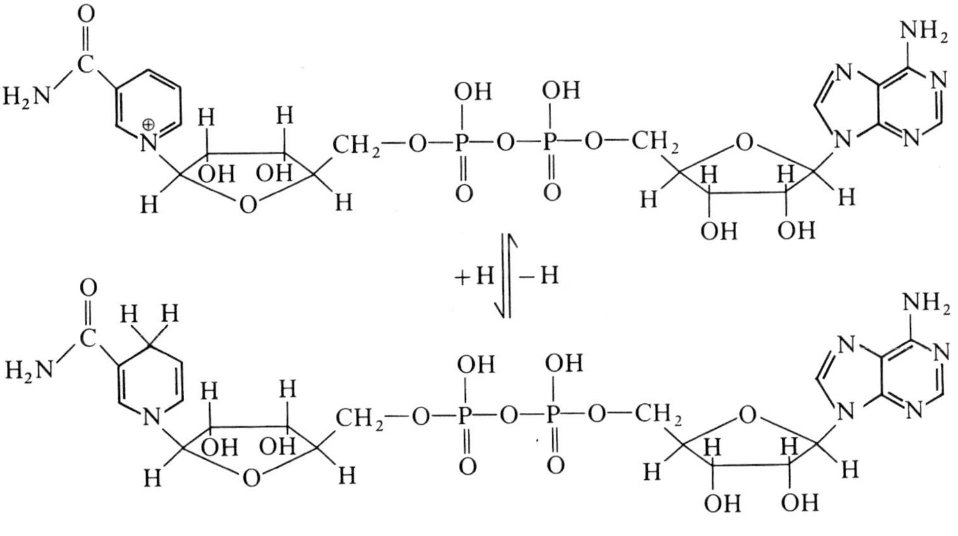

$$\text{NAD} \underset{-\text{H}}{\overset{+\text{H}}{\rightleftharpoons}} \text{NADH}$$

Figure 13 shows that the NAD absorption does not interfere in any way with the NADH absorption band at $\lambda_{max} = 340$ nm. This means that these are ideal conditions for an analytical spectrophotometric NADH determination. In practice, we are dealing with a *final value method*, i.e. the change of the absorbance with time is followed until a final value is reached

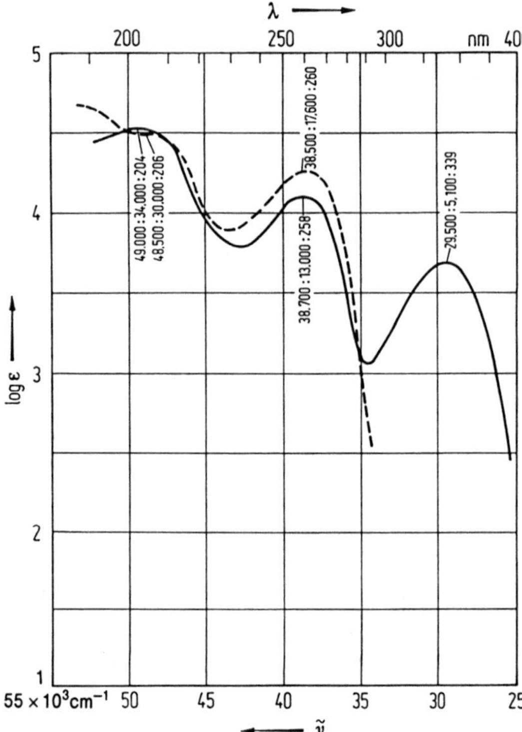

Fig. 13. Absorption spectrum of NAD (– – –), phosphate buffer pH 6.8; and NADH (———), buffer pH 9.0

(see the example in Fig. 15). Subsequently, we can calculate the concentration of this substance, or the enzyme activity, by means of the change ΔA.

Similar reactions which can be measured at 340 nm are given by enzymes such as glutamate-dehydrogenase (GLDH), α-hydroxyl-butyrate-dehydrogenase (HBDH), creatinine-phosphokinase (CPK) and lactate-dehydrogenase (LDH) etc.

Generally, we can describe enzyme-catalyzed reactions using equations for *first, second or pseudo-first* order reactions.

For a first order reaction in its simple form (see Sect. 7.1)

$$S \xrightarrow{\text{enzyme}} B$$

the rate of formation of B is:

$$\frac{d[B]}{dt} = k_B[S] \ . \tag{18}$$

The rate of formation of B is always proportional to the concentration of S present.

The Bouguer-Lambert-Beer law states that the concentration of S is directly proportional to the absorbance of S. Therefore the following applies

$$A_s = \varepsilon_s c_s \cdot d \quad \text{or} \quad c_s = \frac{A_s}{\varepsilon_s d} ,$$

and

$$\frac{d[B]}{dt} = k_b' \cdot A_s^t \quad \text{with} \quad k_B' = k_B (\varepsilon_S d)^{-1} . \tag{19}$$

In the case of the second order reaction we have

$$S + B \xrightarrow{\text{enzyme}} C + D$$

$$\frac{dx}{dt} = k_{CD} \cdot (s_0 - x_S) \cdot (b_0 - x_B) \tag{20}$$

where:

X is the amount reacted,
$s_0 - x_S$ is the specific concentration of S, s_0 = initial concentration of S,
$b_0 - x_B$ is the specific concentration of B, b_0 = initial concentration of B,
k_{CD} is the rate constant for the formation of C+D.

In this case, the rate of formation of C and D is proportional to the product of $[S] \times [B]$.

In enzymatic reactions, there is frequently an excess of one of the reactants. The concentration of such a component, e.g. S, hardly changes from its initial concentration during the course of the reaction which is therefore of pseudo-first order:

$$\frac{dx}{dt} = k_{CD}' (b_0 - x_B) \quad \text{with} \quad k_{CD}' = k_{CD} \cdot [S]_0 . \tag{21}$$

On reintroducing the absorbance we obtain with

$$b = A_B^0 (\varepsilon_B \cdot d)^{-1} = A_B^0 \cdot a \quad \text{and} \quad x_B = A_B^t \cdot a \; ; \quad a = \frac{1}{\varepsilon_B d} ,$$

$$\frac{dx}{dt} = k_{CD}' \cdot a (A_B^0 - A_B^t) = -k_{obs} \cdot A_B^t + \text{const} \tag{22}$$

or

$$\frac{dx}{dt} = k'_{CD} \cdot a A_B^0 \left[1 - \frac{A_B^t}{A_B^0}\right] = k_{obs} \cdot A_B^0 \cdot \left[1 - \frac{A_B^t}{A_B^0}\right] . \tag{22a}$$

The speed of reaction related to the consumption of reactants S and B is again directly proportional to the specific concentration of B.

In contrast to a first order reaction, the velocity constant of a pseudo-first order reaction includes the given concentration of substrate S_0, i.e. $k_{obs} = k_{CD}[S]_0$ a; in this case, the constant a is $(\varepsilon_B d)^{-1}$.

Since we cannot determine the quantity of enzyme present in a reaction directly, the determination is made by measuring its effect, i.e. in enzyme kinetics we measure the catalytic effectiveness on a reaction which can be followed directly or on a subsequent indicator reaction. The available quantity of enzyme is then specified by units which are defined as follows:

The international enzyme unit (U) is the enzyme quantity which changes 1 μmol of substrate in 1 min under optimum standardized conditions:

$$1\,U = \frac{1\,\mu mol}{min} ; \quad 1 \text{ milli unit (mU)} = 10^{-3}\,U .$$

We take the determination of *creatinine-phosphokinase* (CPK) as an example. This enzyme catalyzes the reaction

$$\text{creatine phosphate} + \text{ADP} \xrightleftharpoons{\text{CPK}} \text{creatine} + \text{ATP} \tag{a}$$

The ATP developing during the course of the reaction from left to right is converted by glucose into glucose-6-phosphate (G-6-P) and ADP (reaction b) in the presence of hexokinase (HK). Subsequently, G-6-P is converted by NAD into 6-phosphogluconate and NADH in the presence of glucose-6-phosphate-dehydrogenase (reaction c):

$$\text{ATP} + \text{glucose} \xrightarrow{\text{HK}} \text{G-6-P} + \text{ADP} \tag{b}$$

$$\text{G-6-P} + \text{NAD} \xrightarrow{\text{G-6-PDH}} \text{NADH} + \text{6-phosphogluconate} . \tag{c}$$

The third reaction is the most valuable and important one for spectroscopy. The reactions are strictly stoichiometric and the NADH developing in reaction (c) corresponds to the progress of reaction (a), i.e. the enzyme activity.

In order to carry out the reaction, we have to select appropriate concentrations of creatine phosphate, ADP, glucose and NAD and, after adding enzyme CPK, we measure the increase of absorbance A at $\lambda = 340$ nm as a function of time in this pseudo-first order reaction.

Figure 14 shows data from Long [132] for the course of a reaction with time measured at 303 K. The temperature must be kept constant within

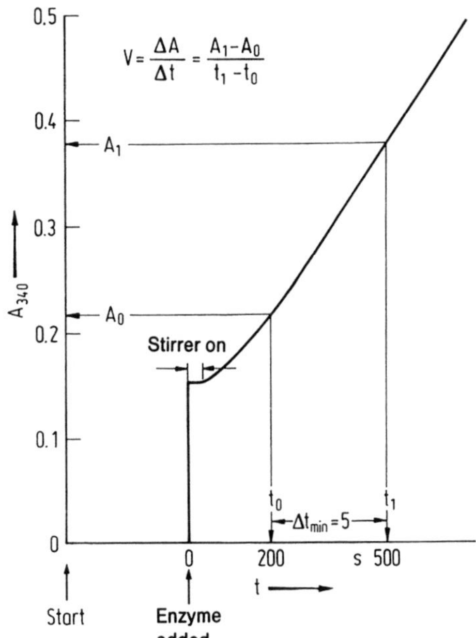

Fig. 14. Absorbance v. time for enzyme kinetic reactions (a), (b) and (c) discussed in the text [132]

± 0.02 degrees. Between points t_0 and t_1 300 s = 5 min elapsed. The absorbances were $A_0 = 0.216$ and $A_1 = 0.376$ and from these we obtain the change of the absorbance per minute as:

$$\frac{\Delta A}{\Delta t} = \frac{0.376 - 0.216}{5} = \frac{0.150}{5} = 0.03\ \mathrm{A/min}\ .$$

With the above value of $\Delta A/\Delta t$, $d = 1$ cm and $\varepsilon_{340} = 6.22 \times 10^3\ \mathrm{mol^{-1}\ cm^{-1}}$ for NADH we have, since $c = A/\varepsilon d$ and $a = (\varepsilon d)^{-1}$:

$$\frac{\Delta c}{\min} = \frac{\Delta A}{\varepsilon \cdot d} = \frac{0.03}{6.22 \cdot 10^3 \cdot 1} = 0.482 \times 10^{-6}\ \mathrm{mol\ l^{-1}\ min^{-1}}\ .$$

0.1 mg enzyme substrate were added to a sample volume of 3 ml in the cuvette in this test. Taking account of the dilution factor 3.1/0.1 and the conversion of l to ml, we finally obtain the enzyme activity of CPK as

$$0.01494\ \mathrm{U\ ml^{-1}\ min^{-1}} \triangleq 14.94\ \mathrm{mU\ ml^{-1}\ min^{-1}}\ .$$

This example can be applied to other enzymes if the specific reaction conditions are taken into account. Appropriate experimental instructions can be found in [119] and [133].

Mattenheimer [134] has briefly discussed the theory of the enzymatic test. Cornish-Bowden [135] and Bergmeyer [136] have given detailed accounts of the principles of enzymatic analysis and enzyme kinetics.

An important application of enzyme kinetics is based on the fact that enzyme activity can be influenced by substrates. This can be understood by assuming the *formation of an enzyme substrate complex* (ES) between enzyme (E) and substrate (S); this complex is capable of reacting further to form product (P) and the enzyme. The reaction equations can be written as follows:

$$E+S \underset{k_{-1}}{\overset{k_1}{\rightleftharpoons}} ES \ , \quad v_1 = k_1[E]\cdot[S] \ , \tag{23}$$

$$ES \xrightarrow{k_2} P+E \ , \quad v_2 = k_2[ES] \ . \tag{23a}$$

The conditions for the Bodenstein approximation (stationary state) are met if the enzyme concentration is considerably smaller than the sum of the substrate and product concentrations [137]. In this case, the following applies to the rates of a reaction:

$$k_1[E]\cdot[S]-(k_{-1}+k_2)[ES]=0 \ . \tag{24}$$

With $[E]_0 = [E]+[ES]$ as total concentration of the enzyme, the concentration of the enzyme-substrate complex is obtained from Eq. (24).

$$[ES] = \frac{k_1[E]_0[S]}{k_{-1}+k_2+k_1[S]} \ . \tag{25}$$

The rate of formation of product P is then

$$v_2 = \frac{d[P]}{dt} = k_2[ES] = \frac{k_1 k_2 [E]_0 [S]}{k_{-1}+k_2+k_1[S]} \ . \tag{26}$$

Michaelis and Menten [138] derived an analogous relationship in 1913. They assumed that the enzyme-substrate complex is formed in a rapid, reversible process, i.e.

$$k_{-1} \gg k_2$$

$$v = \frac{k_1 k_2 [E]_0 [S]}{k_{-1}+k_1[S]} \ . \tag{27}$$

However, this assumption is not always correct as Chance has shown [139]. In enzyme kinetics, Eqs. (23) and (24) are used in the form:

$$v = \frac{k_2 [E]_0 [S]}{\frac{k_{-1} + k_2}{k_1} + [S]} = \frac{k_2 [E]_0 [S]}{K_M + [S]} \tag{28}$$

or

$$v = \frac{k_2 [E]_0 [S]}{\frac{k_{-1}}{k_1} + [S]} = \frac{k_2 [E]_0 [S]}{K'_M + [S]} \tag{28a}$$

where the Michaelis constant $K_M = \frac{k_{-1} + k_2}{k_1}$ or $K'_M = \frac{k_{-1}}{k_1}$, is equal to the substrate concentration at which the rate of a reaction has been reduced by half of its initial maximum value. Cornish-Bowden has given a detailed discussion of these relationships for enzyme-catalyzed reactions [135].

Pautler and Jackson have reported a UV-spectroscopic investigation of a three-stage process where a complex is formed between *peptides and proteinase* [140]. Moody and Heisz have described the *Emit homogenous enzyme immunoassay* from the Syva company in an application to therapeutical *drug monitoring* (EMIT-TDM) by means of UV-VIS spectroscopy [141].

Commonly used methods of enzymatic analysis assume that specific organic compounds are oxidized or reduced by the catalytic action of an enzyme where the coenzymes NAD and NADH usually participate.

For example, *alcohol* is oxidized to acetaldehyde by NAD if the alcohol dehydrogenase enzyme is present:

$$C_2H_5OH + NAD^+ \xrightleftharpoons{ADH} CH_3CHO + NADH + H^+ \ . \tag{a}$$

The equilibrium lies on the side of ethanol and NAD in this reaction. Equilibrium (a) can be moved to the right by an alkaline environment or by removing acetaldehyde. This is oxidized quantitatively to acetic acid if aldehyde-dehydrogenase (AL-DH) is available:

$$CH_3CHO + NAD^+ + H_2O \xrightarrow{Al\text{-}DH} CH_3COOH + NADH + H^+ \ . \tag{b}$$

These two reactions are carried out in tandem so that 2 moles of NADH are obtained for 1 mole of ethanol in the overall reaction. Since the absorption maximum of NADH lies at $\lambda = 340$ nm, the reaction can be followed quantitatively if the final value of the absorbance in the solution is determined.

The concentration of the substance under investigation is:

$$c = \frac{V\,MW}{\varepsilon \cdot d \cdot v \cdot n \cdot 1000} \cdot \Delta A \ \text{in g}\,l^{-1} \ . \tag{29}$$

where

V is the test volume in ml;
v is the sample volume in ml;
MW is the molecular weight of the substance under investigation;
d is the pathlength of the cuvette in cm;
n is the stoichiometric coefficient referred to the substance measured directly; n = 2 in the above example;
ε is the extinction coefficient in $l\,mmol^{-1}\,cm^{-1}$
(If we include the factor 1000 in ε, then we obtain ε as molar decadic extinction coefficient in $l\,mol^{-1}\,cm^{-1}$).

The determination of *β-D-glucose* provides a useful example [142, 143]. D-Glucose exists in solution in the two anomeric forms α- and β-D-glucose in the ratio of 1 : 2. The anomers are in equilibrium by mutarotation.

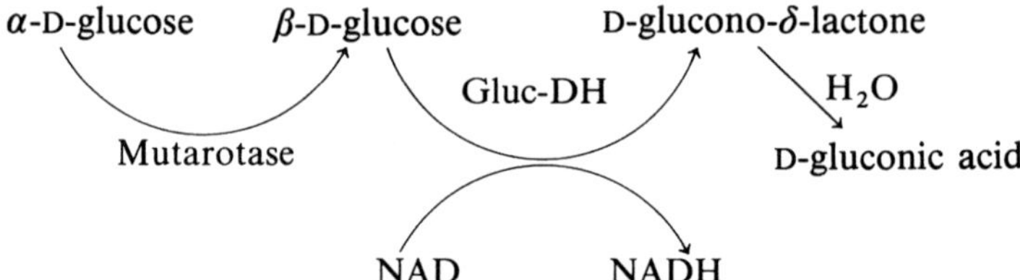

The oxidation of β-D-glucose to D-glucono-δ-lactone under the catalytic effect of the enzyme glucose dehydrogenase with the involvement of co-

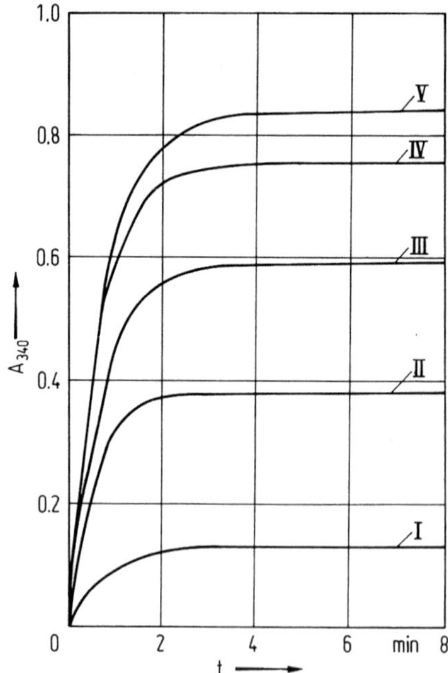

Fig. 15. β-D-glucose; final absorbance value determination for different specific concentrations

Table 12. Summary of enzymatic analyses monitored by UV photometry. Enzymes which recur in the table are indicated by their abbreviated names only

Compound	Enzymes involved: stages	Analyt. react.
Acetaldehyde	Aldehyde dehydrogenase (Al-DH); single-stage	NAD→NADH
Acetic acid	Acetyl-CoA synthetase (ACS); citrate synthase (CS); malate dehydrogenase (MDH); three-stage	NAD→NADH
L-asparagine	Asparagenase; glutamate-oxalacetate transaminase (GOT); MDH; three-stage	NAD←NADH
Citric acid	Citrate lyase (CL); MDH; lactate dehydrogenase (LDH); three-stage	NAD←NADH
Creatine	Creatinase, creatine kinase (CK)	NAD←NADH
Creatinine	Pyruvate kinase (PK); LDH; four-stage	
Ethanol	Alcohol dehydrogenase (ADH); Al-DH; two-stage	NAD→NADH
Formic acid	Formate dehydrogenase (FDH); single-stage	NAD→NADH
D-gluconic acid	Gluconate kinase; 6-phosphogluconate dehydrogenase (6-PGDH); two-stage	NAD→NADH
Glucose Fructose	Hexokinase (HK); glucose-6-phosphate-dehydrogenase (G6P-DH); two-stage	NAD→NADH
L-glutamic acid	Glutamate dehydrogenase (GlDH) iodonitrotetra-zoliumchloride (INT+diaphorase); two-stage	NAD→NADH←→formazon
Glycerine	Glycerokinase (GK); PK; LDH; three-stage	NAD←NADH
Guanosine-5′-monophosphate	Guanosine-5′ monophosphate kinase (G-5-MPK); PK; L-LDH; three-stage	NAD←NADH
Isocitric acid	Isocitrate dehydrogenase (ICDH); single-stage	NAD→NADH
L-lactic acid	L-LDH; GPT; two-stage	NAD→NADH
D-lactic acid	D-LDH; GPT; two-stage	NAD→NADH
Lactose/galactose	β-galactosidase; β-galactose dehydrogenase (Gal-DH); two-stage	NAD→NADH
Lecithin	Phospholipase C; alkaline phosphatase (AP); choline kinase; PK; LDH; five-stage	NAD←NADH
L-malic acid	L-MDH; GOT; two-stage	NAD→NADH
Maltose	α-glucosidase (maltase); hexokinase (HK) G-6-P-DH; three-stage	NAD→NADH
Pyruvic acid (pyruvate)	LDH; single-stage; see succinic acid, third stage	NAD←NADH
Raffinose	α-galactosidase; Gal-DH; two-stage	NAD→NADH
Saccharose	Enzymatic hydrolysis, then see glucose	NAD→NADH
D-sorbitol	Sorbitol dehydrogenase (SDH); single-stage	NAD→NADH
D-sorbitol/xylitol	SDH; diaphorase; two-stage/see L-glutamic acid	NAD→NADH→→formazon
Starch	Amyloglucosidase (AGS); HK; G6P-DH; three-stage	NAD→NADH
Succinic acid	Succinyl-CoA-synthetase (SCS); pyruvate kinase (PK); LDH; three-stage	NAD←NADH
Triglyceride	Lipase+esterase; GK; PK; LDH; four-stage	NAD←NADH
Urea/ammonia	Urease, Gl-DH; two-stage	NAD←NADH

enzyme NAD as H-acceptor is the decisive reaction, i.e. the course of the whole reaction can be followed by monitoring the increase in the absorbance maximum of NADH ($\lambda_{max} = 340$ nm) to its final value.

Since gluc-DH is strictly specific for β-D-glucose, the self-mutarotation of glucose is the rate-determining step when establishing the final value of the NADH absorbance. Thus, a constant value is not obtained in less than ca. 15 – 20 min. If the mutarotation is catalyzed by mutarotase, a constant final value is obtained after 5 – 8 min. Figure 15 shows the absorbance increases at $\lambda = 340$ nm for five initial concentrations of D-glucose taken directly from the recording; the final values are obtained after 5 – 6 min.

The glucose concentration is calculated in g glucose per 1 sample solution as shown in Eq. (29). Glucose determination in body fluids (blood, serum, plasma, liquor, urine) requires additional operating instructions e.g. for the removal of albumen [143].

Table 12 contains a summary of compounds for which a quantitative determination via the final absorbance value of the participating NADH may be made.

These compounds are often substances found in foods. Therefore, *enzymatic food analysis* is of particular importance. Boehringer/Mannheim have published a collection of experimental methods [142]. Henniger has provided a further discussion [144]. Krüger and Nordmann have dealt with the enzymatic determination of some of the above compounds in beer and apple juice [145].

4.2 Multicomponent Analysis

4.2.1 Basic Equations

Equation (14a) represents the basic equation for multi-component analysis:

$$A_i = \{\varepsilon_{i1} \cdot c_1 + \varepsilon_{i2} \cdot c_2 + \varepsilon_{i3} c_3 + \ldots\} \, d = d \sum_{j=1}^{n} \varepsilon_{ij} \cdot c_j \; . \tag{14a}$$

Index i refers to the wavenumber or wavelength and the second index j to the component (j = 1 to n).

Since each individual extinction coefficient ε_{ij} is a characteristic function of λ or $\tilde{\nu}$ this also applies to the mixture. Therefore, for a system of equations with n unknowns, we obtain the corresponding number of determining equations by measuring at n wavelengths:

$$\begin{aligned}
A_1 &= d \sum_{j=1}^{n} \varepsilon_{1j} c_j \, , & D_1 &= \sum_{j=1}^{n} \varepsilon_{1j} c_j \, , \\
A_2 &= d \sum_{j=1}^{n} \varepsilon_{2j} c_j & D_2 &= \sum_{i=j}^{n} \varepsilon_{2j} c_j \\
&\vdots & &\vdots \\
A_n &= d \sum_{j=1}^{n} \varepsilon_{nj} c_j \, , & D_n &= \sum_{j=1}^{n} \varepsilon_{nj} c_j \; .
\end{aligned} \tag{14b}$$

$A_i/d = D_i$ is defined here as *optical density.* Equation (14b) is the mathematical foundation of multicomponent analysis which permits the determination of a concentration in solution without having to interfere with the system, provided that we know the components and their extinction coefficients at the n different wavelengths.

When carrying out a multicomponent analysis we must always be aware of several premises and conditions:

1. The Bouguer-Lambert-Beer law must be valid, i.e. absorbances must be additive over the concentration range in question, i.e. components must not interact.
2. The greater the similarity between the spectra of individual components the more difficult and inaccurate the analysis. The choice of optimum wavelengths for the spectral analysis plays an important role.
3. Interaction with the solvent must be excluded.
4. Values taken from the steep flanks of spectral bands should not be included in the analysis since an error in setting the wavelength causes a large error in absorbance, i.e. $dA/d\lambda$ is very large in this case. In contrast, $dA/d\lambda$ is small in the region of the absorption maxima, absorption minima and pronounced shoulders.
5. Very large or very small absorbances should be avoided since photometric accuracy is inferior in both cases.
6. Impurities absorbing in the same region can lead to considerable errors when measuring absorbances.
7. The resolution of a spectrophotometer (slit program) should be set in such a way that the bands are optimally resolved. The measurement of the mixture must take place under the same instrumental conditions as that of the components.

An analysis of a *dual-component system* is frequently straightforward since we can select wavelengths for an analysis which, in many cases, meet condition 2 and where, under certain circumstances, only one component will absorb at a certain wavelength. In this case, we have the photometric conditions for the determination of a single substance. Molch, König and Than [146] have developed graphical methods of evaluation for the simultaneous spectrophotometric determination of two-component systems. The authors base their evaluation of $c_A - c_B$ diagrams on grids or nomograms. This method has been developed especially for the determination of complexes and takes account of the fact that the reagent is also present in the solution.

Harker et al. have reviewed and optimized the above mentioned conditions for a *three-component analysis* [147] using a mixture of three isomeric cresols. They obtained an accuracy of 1% – 5%.

The complete mathematical formulation of a three-component system, with the introduction of the optical density, is given by:

$D = A \cdot d^{-1}$ (d = pathlength)

$$\begin{aligned} D_1 &= \varepsilon_{11} c_1 + \varepsilon_{12} c_2 + \varepsilon_{13} c_3 \, , \\ D_2 &= \varepsilon_{21} c_1 + \varepsilon_{22} c_2 + \varepsilon_{23} c_3 \, , \\ D_3 &= \varepsilon_{31} c_1 + \varepsilon_{32} c_2 + \varepsilon_{33} c_3 \, . \end{aligned} \tag{27/1}$$

Matrix notation greatly simplifies matters and every equation of (27/1) can be written in the form:

$$D_1 = [\varepsilon_{11} \varepsilon_{12} \varepsilon_{13}] \begin{pmatrix} c_1 \\ c_2 \\ c_3 \end{pmatrix} = \sum_j \varepsilon_{1j} c_j = E_1 \cdot C \, . \tag{28/1}$$

D_2 and D_3 can be similarly formulated and this combination yields Eq. (29/1):

$$\begin{pmatrix} D_1 \\ D_2 \\ D_3 \end{pmatrix} = \begin{pmatrix} \varepsilon_{11} & \varepsilon_{12} & \varepsilon_{13} \\ \varepsilon_{21} & \varepsilon_{22} & \varepsilon_{23} \\ \varepsilon_{31} & \varepsilon_{32} & \varepsilon_{33} \end{pmatrix} \begin{pmatrix} c_1 \\ c_2 \\ c_3 \end{pmatrix} \tag{29/1}$$

or

$$D = E \cdot C \, . \tag{29/1 a}$$

If only one or two unknown ternary mixtures have to be analyzed then the equation system (29/1) can be solved by means of Cramer's rule.

The values of D_i measured optically are substituted for the extinction coefficients in the first column of the E-matrix. The determinant of the new matrix is then divided by that of the original E-matrix to obtain the value of c_1:

$$c_1 = \frac{\begin{vmatrix} D_1 & \varepsilon_{12} & \varepsilon_{13} \\ D_2 & \varepsilon_{22} & \varepsilon_{23} \\ D_3 & \varepsilon_{32} & \varepsilon_{33} \end{vmatrix}}{\begin{vmatrix} \varepsilon_{11} & \varepsilon_{12} & \varepsilon_{13} \\ \varepsilon_{21} & \varepsilon_{22} & \varepsilon_{23} \\ \varepsilon_{31} & \varepsilon_{32} & \varepsilon_{33} \end{vmatrix}} \, . \tag{30}$$

The concentrations of the other components, c_2 and c_3, are obtained in the same manner. D is substituted in the j-th column of E and this new determinant is again divided by determinant E in order to obtain c_j.

If we have to deal with many unknown ternary or higher multicomponent systems which always contain the same components – a problem frequently occurring in routine analysis – then it is more appropriate to form the inverse matrix of E.

Multiplying Eq. (29/1) from the left by E^{-1} we obtain

$$E^{-1}D = E^{-1}EC = C \tag{31}$$

or

$$c_j = \sum (\varepsilon_{ji}^{-1}) \cdot D_1 \ , \tag{32}$$

i.e. if matrix E^{-1} is known, each concentration of c_j in a three-component system is obtained by multiplying three pairs of numbers and adding up the products.

The formation of an inverse matrix, also called a reciprocal matrix, is complicated and its computation is time-consuming. If the determinant of the numerator in the first column of Eq. (30) is expanded by means of Laplace's expansion a linear equation is again obtained in the form:

$$c_1 = \frac{\alpha_{11}}{|E|} D_1 + \frac{\alpha_{21}}{|E|} D_2 + \frac{\alpha_{31}}{|E|} D_3 \tag{33}$$

and correspondingly for c_2 and c_3:

$$c_2 = \frac{\alpha_{12}}{|E|} D_1 + \frac{\alpha_{22}}{|E|} D_2 + \frac{\alpha_{32}}{|E|} D_3 \ ,$$

$$c_3 = \frac{\alpha_{13}}{|E|} D_1 + \frac{\alpha_{23}}{|E|} D_2 + \frac{\alpha_{33}}{|E|} D_3 \ .$$

The α_{ij} are the cofactors of the original determinant E which appears in the denominator in Eq. (30).

The coefficients of these three equations again form a matrix which we call E^{-1} since it is the inverse of matrix E

$$E^{-1} = \begin{pmatrix} \frac{\alpha_{11}}{|E|} & \frac{\alpha_{21}}{|E|} & \frac{\alpha_{31}}{|E|} \\ \frac{\alpha_{12}}{|E|} & \frac{\alpha_{22}}{|E|} & \frac{\alpha_{32}}{|E|} \\ \frac{\alpha_{13}}{|E|} & \frac{\alpha_{23}}{|E|} & \frac{\alpha_{33}}{|E|} \end{pmatrix} \ . \tag{34}$$

The cofactors are derived from the original determinant E by setting up sub-determinants. However, note must be taken of the algebraic sign. In the case of our triple-row determinant, nine cofactors must be determined. It can be easily seen that this arithmetic method quickly becomes unmanageable if more than three components are present. Thus, this technique can only be used if a computer is available. We usually work with overdetermined inhomogenous equation systems as far as multicomponent analyses are concerned. Starting from Eq. (14b), the problem can be formulated in matrix notation as follows:

$$\begin{pmatrix} D_1 \\ D_2 \\ \vdots \\ D_m \end{pmatrix} = \begin{pmatrix} \varepsilon_{11} & \varepsilon_{12} & \cdots & \varepsilon_{1n} \\ \varepsilon_{21} & \varepsilon_{22} & \cdots & \varepsilon_{2n} \\ \cdots & \cdots & \cdots & \cdots \\ \varepsilon_{m1} & \cdots & \cdots & \varepsilon_{mn} \end{pmatrix} \times \begin{pmatrix} c_1 \\ c_2 \\ \vdots \\ c_n \end{pmatrix} . \tag{35}$$

Elements D_i of matrix D are the optical densities measured at m wavelengths. Elements ε_{ij} of matrix E are the extinction coefficients of the n components at m wavelengths and elements c_j of matrix C are the concentrations of the n components. Hence we can write Eq. (35) like Eq. (29/1 a). If $m = n$, i.e. if the number of measurements at different wavelengths is equal to that of the components, we obtain the square matrix E whose inverse E^{-1} can be written as above, provided E is known. Thus, the elements of matrix C can also be obtained.

For $m < n$, i.e. fewer wavelengths than unknown factors, a solution cannot be obtained for matrix C. For $m > n$, i.e. more wavelengths than unknown factors, we are dealing with an overdetermined equation system and obtain a multitude of solutions for matrix C when using different sets of equations. Since the elements of matrix E and D include errors of measurement it is usually not possible to meet the requirements of Eqs. (29/1) or (35) exactly.

However, an optimum solution for n unknown factors can be obtained by the *method of least squares*. Solving this problem mathematically gives Eq. (36) which is analogous to Eq. (29/1 a).

$$D' = E' \cdot C \tag{36}$$

or

$$E'^{-1} D' = E'^{-1} \cdot E' \cdot C = C \tag{36a}$$

with the abbreviations $D' = DE^{\perp}$ and $E' = EE^{\perp}$.

Matrix $E^{\perp}$ is the matrix transpose of E which can be formed from elements ε_{ij} as shown in Eqs. (37, 37 a):

$$E = \begin{pmatrix} \varepsilon_{11} & \varepsilon_{12} & \cdots & \varepsilon_{1n} \\ \varepsilon_{21} & \varepsilon_{22} & \cdots & \varepsilon_{2n} \\ \cdots & \cdots & \cdots & \cdots \\ \varepsilon_{m1} & \varepsilon_{m2} & & \varepsilon_{mn} \end{pmatrix} , \tag{37}$$

$$E^{\perp} = \begin{pmatrix} \varepsilon_{11} & \varepsilon_{21} & \cdots & \varepsilon_{m1} \\ \varepsilon_{12} & \varepsilon_{22} & \cdots & \varepsilon_{m2} \\ \vdots & \vdots & \cdots & \vdots \\ \varepsilon_{1n} & \varepsilon_{2n} & & \varepsilon_{mn} \end{pmatrix} . \tag{37a}$$

Matrix E′ is square with dimension $n \times n$ because it results from multiplying the $n \times m$ matrix E with the $m \times n$ matrix $E^{\perp}$. In contrast, matrix D′ is of the dimension $n \times 1$. We obtain the required optimized solutions for the n unknown concentrations from Eq. (36a), which corresponds to Eq. (31) [148, 149].

Sternberg et al. [148] applied this method to the analysis of the *photoisomerization of ergosterol.* After irradiation, some ergosterol remains together with lumisterol, tachysterol, calciferol and precalciferol, i.e. we have a five-component system. The accuracy of this method was discussed in detail using test compositions of the five components and a variety of selected wavelength combinations.

Zscheile et al. also analysed a *four-component system* consisting of the RNA constituents adenylic, cytidylic, guanylic and uridylic acid by this method [150].

The question, of how far it is practical to overdetermine a system of inhomogeneous equations in order to achieve adequately accurate results, was investigated comprehensively by Herschberg [151] who used a modified technique. He concluded that 20–40 wavelengths are sufficient for a maximum of six components. In order to minimise errors of measurement Herschberg et al. measured the extinctions of pure components at the same time as those of the compositions to be analyzed. They discussed the analysis of the following multicomponent systems [152, 153, 154].

a) Benzene, toluene, ethylbenzene, *o*-xylene, *m*-xylene, *p*-xylene (in isooctane);
b) Benzene sulfonic acid, *p*-tert. butyl benzene sulfonic acid, *m*-tert. butyl benzene sulfonic acid (in H_2SO_4 83.6%);
c) Benzene sulfonic acid, *o*-toluene sulfonic acid, *m*-toluene sulfonic acid, *p*-toluene sulfonic acid (in water);
d) *o*-toluene sulfonic acid, *p*-toluene sulfonic acid, *o*-xylene-3-sulfonic acid, *o*-xylene-4-sulfonic acid (in H_2SO_4 77.8%);
e) *n*-propylbenzene, *iso*-propylbenzene (in CCl_4).

A six-component analysis is therefore possible.

The *three-component systems* methylphenyl-sulfide, -sulfoxide and -sulfone, methylbutyl-, methyloctyl- and methyldodecyl-sulfide were investigated in the same way [155]. The first system gave useful values using only ten equations since the absorption spectra differed considerably in the region of 36000–44000 cm^{-1}. The spectra of the set of homologous sulfides hardly differ in the region 38000–47000 cm^{-1}. Acceptable results could not be obtained with less than 30 wavelengths.

In these examples the decision about the number of overdetermining equations required was made on purely empirical grounds. However, it is clear that the influence of random errors increases with an increasing number of test data. Therefore, the two effects can compete. Sustek has investigated this problem on a five-component system which was simulated by computer [156]. In total, he set a maximum of 32 analytical positions (wavelengths) from which he selected groups of 5, 10, 15, 20, 25 and 32 in order to record the influence of an increasing number of analytical wavelengths on the accuracy of the result. The dependence of the results upon the random errors of measurement was investigated in more detail on three model systems. For model system 1, the test data (optical densities, matrix D and extinction coefficients, matrix E) were regarded as error-free. In system 2, errors of measurement were pre-set at each analytical wavelength and in model system 3 the errors of measurement were statistically distributed about the analytical wavelengths. Model system 1 showed, as expected, that a decrease of the standard deviation results from an increase in the number of analytical wavelengths. Model system 3 shows how both effects compensate. However, the negative influence of random errors becomes noticable. In model system 2 the standard deviation is seen to decrease initially with an increasing number of analytical wavelengths; subsequently, the influence of random errors increasingly impairs the result. As a result of the investigations of these model systems, it can be said that where there are overlapping absorption bands a triply or quadruply overdetermined system forms an optimum basis for a multicomponent analysis. Account must be taken of the magnitude of random errors of measurement.

Leggett and McBryde have dealt with a case where one or even several *equilibria* are present in a multicomponent analysis for which an appropriate program (SQUAD) was written [157]. This method can be used for determining pk_a values, hydrolysis constants and multicomponent equilibria. In further work the program was extended by inclusion of the algorithms for least squares analysis, see Eqs. (36) and (36a) [158].

As examples, *mixtures of indicator dyes* such as bromcresol green, phenol red and methyl orange were analyzed and their associated pk_a values in the mixture were determined.

Fahr and Schmid have described a multicomponent analysis and its application to *nucleic acid components* [159]. It is of interest that we are dealing here with a system which is also varying with time.

With the development of modern UV-VIS spectrophotometers fitted with microprocessors and/or microcomputers, multicomponent analysis has become straightforward in many applications since suitable programs are provided as routine software for many instruments. In many cases, automated instruments can be connected to a remote microcomputer.

James [160] and Long and Greig [161] give examples of multicomponent analysis with UV-VIS spectrophotometers and microcomputers. Jochum et al. have dealt with the problem of error propagation and the optimization of a multicomponent analysis [162] based on the *generalized standard addition method* (GSAM) [163].

In addition to the evaluation of overdetermined systems by means of matrix calculations incorporating algorithms for minimizing the error squares, other techniques have been proposed for multicomponent analysis. One of these is the *iteration technique* [164]. Another method designed for two-component systems is a multipoint technique which allows a graphical evaluation [165]. Of course, none of the techniques are error free.

A detailed investigation of the sources of errors in multicomponent analyses has been carried out by Bergmann [166].

4.2.2 An Example of a Multicomponent Analysis

Numerical evaluation according to Eq. (30) and (36a) will be described briefly for the three-component system *o*-, *m*- and *p*-nitrophenol in 0.1 *m* NaOH.

Table of the experimental data D_i and extinction coefficients ε_{ij}: i = 1 to 15; j = 1 to 3; *o*NP = *o*-nitrophenol; *m*NP = *m*-nitrophenol; *p*NP = *p*-nitrophenol:

i	$\tilde{\nu}_i$ $\|cm^{-1}\|$	D_i	$\varepsilon_{i,pNP}$	$\varepsilon_{i,oNP}$	$\varepsilon_{i,mNP}$
1	23000	0.695	8290	4107	917
2	24000	0.958	15635	4675	1284
3	25000	1.032	18782	4324	1486
4	26000	0.915	16580	3382	1490
5	27000	0.708	12067	2356	1335
6	28000	0.530	7975	1546	1147
7	29000	0.405	5140	1075	1018
8	30000	0.360	3253	882	1069
9	31000	0.440	1994	966	1560
10	32000	0.675	1322	1418	2523
11	33000	0.990	1028	1425	3601
12	34000	1.200	1028	3358	4211
13	35000	1.240	1091	4131	4014
14	36000	1.120	1406	4192	3454
15	37000	1.160	2204	3624	4083

The evaluation via Eq. (30) with values for i = 1 − 3 yields the expression:

$$c_{pNP} = \frac{\begin{vmatrix} 0.695 & 4107 & 918 \\ 0.958 & 4675 & 1284 \\ 1.032 & 4324 & 1486 \end{vmatrix}}{\begin{vmatrix} 8290 & 4107 & 917 \\ 15635 & 4675 & 1284 \\ 18782 & 4324 & 1486 \end{vmatrix}} = \frac{-6.068 \times 10^4}{-3.334 \times 10^9} = \underline{\underline{1.817 \times 10^{-5}\,m}} \; .$$

We obtain the two other concentrations analogously:

$$c_{m\text{NP}} = 2.395\times10^{-4}\,m \quad \text{and} \quad c_{o\text{NP}} = 8.184\times10^{-5}\,m\ ;$$

the concentrations in the mixture were set at:

$$c_{p\text{NP}} = 1.906\times10^{-5}\,m\ ;\ c_{m\text{NP}} = 2.18\times10^{-4}\,m\ ;\ c_{o\text{NP}} = 8.28\times10^{-5}\,m\ .$$

The deviations from the known values are accordingly ~4% for *p*-nitrophenol, ~10% for *m*-nitrophenol and ~1.2% for *o*-nitrophenol.

For the evaluation by Eq. (36), the values of the first ten (i = 1 – 10) wavenumbers listed in the table were used initially.

The measured data in column D_i form matrix D. The extinction coefficients $\varepsilon_{1,1}$ to $\varepsilon_{10,3}$ constitute the three columns of matrix E which is of dimension 3×10.

From this we obtain the transposed matrix $E^{\perp}$ of dimension 10×3:

$$1)\ E^{\perp} = \begin{pmatrix} 8290 & 15635 & 18782 & 16580 & 12067 & 7975 & 5140 & 3253 & 1994 & 1322 \\ 4107 & 4675 & 4324 & 3382 & 2356 & 1546 & 1075 & 882 & 966 & 1418 \\ 917 & 1284 & 1486 & 1490 & 1335 & 1147 & 1018 & 1069 & 1560 & 2523 \end{pmatrix} .$$

$E \times E^{\perp} = E'$, the 3×3 matrix:

$$2)\ E' = \begin{pmatrix} 119275692 & 297382231 & 120704315 \\ 297382231 & 81676255 & 33273767 \\ 120704315 & 33273767 & 20993889 \end{pmatrix} .$$

Matrix $D' = D \times E^{\perp}$ is:

$$3)\ D' = \begin{pmatrix} 73086.28 \\ 19512.426 \\ 9503.934 \end{pmatrix} .$$

When calculating concentrations according to Eq. (36a) we also require the inverse matrix E'^{-1} which is derived from E':

$$4)\ E'^{-1} = \begin{pmatrix} 9.09223\times10^{-9} & -3.33269\times10^{-8} & 5.43994\times10^{-10} \\ -3.33263\times10^{-8} & 1.56707\times10^{-7} & -5.67603\times10^{-8} \\ 5.43994\times10^{-10} & -5.67603\times10^{-8} & 1.34466\times10^{-7} \end{pmatrix} .$$

Thus, using Eq. (36a) we obtain the values in mol/l listed in column 2 of the following table:

	Using Eq. (36a) i = 10	Given	deviation	Using Eq. (36a) i = 15
p-Nitrophenol	1.941×10^{-5}	1.906×10^{-5}	1.8%	2.001×10^{-5} (4.3%)
m-Nitrophenol	2.102×10^{-4}	2.180×10^{-4}	4%	2.212×10^{-4} (1.5%)
o-Nitrophenol	8.260×10^{-5}	8.280×10^{-5}	1%	7.554×10^{-5} (9%)

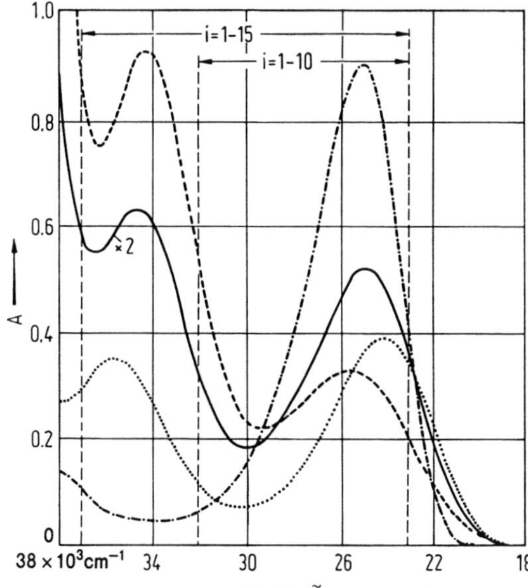

Fig. 16. Absorption spectra of the three isomeric nitrophenols and a mixture in 0.1 m NaOH as an example of a multicomponent analysis, see Sect. 4.2.2. *p*-NP (–·–·–), *m*-NP (– – –), *o*-NP (. . .). mixture (———)

In order to check if the result can be improved by supplementary analytical data points, the values for five more wavenumbers (i = 11 – 15) were included. The last column of the above table shows that an improved result is achieved only with *m*-nitrophenol whilst a deterioration is observed in the case of the other nitrophenols. An explanation for this behaviour can be found by examination of the spectra of the pure compounds in 0.1 m NaOH used to determine the extinction coefficients ε_{ij} for matrix E. For *p*-nitrophenol, the absorbances in this supplementary wavenumber region between 33000 cm^{-1} and 37000 cm^{-1} lie below A = 0.06.

The extinction coefficients obtained from this therefore carry substantially greater errors than in the region of $\tilde{\nu}_1 \rightarrow \tilde{\nu}_{10}$ (23000 – 32000 cm^{-1}).

m-Nitrophenol shows an intense absorption band at 34300 cm^{-1} in the region of 33000 – 37000 cm^{-1}. Accordingly, the A-values can be read with adequate accuracy from the recorded curve and the $\varepsilon_{i,mNP}$ values thus established can be treated as very reliable.

Figure 16 shows the absorption spectra of the three phenols in 0.1 m NaOH in the region of $18000 \leq \tilde{\nu}_i \leq 38000$ cm^{-1} and also the superimposed spectrum of the mixture of all three nitrophenols in the same wavenumber region. This example demonstrates the importance of the selection of the analytical wavelengths for a multicomponent analysis. All spectra were measured consecutively with a Perkin-Elmer UV 320 dual-beam spectrophotometer which, to judge from the results obtained, provides an excellent reproducibility of the wavenumber setting.

Further reliability in the results can be achieved by connecting a printer since reading the absorbance values from the recorded spectra is relatively inaccurate. The data were evaluated by means of a program written for the computer HP9830.

4.3 Identification and Structure Determination

The application of UV-VIS spectroscopy in identification and structure determination is based on the fact that the electronic ground and excited states of molecules depend upon the number of electrons, the structure and geometry and also the symmetry. These correlations have been fully investigated theoretically, particularly in the case of unsaturated organic compounds [167–171]. In addition to structural influences, the effects of substituents and the presence of heteroatoms must be considered.

Some typical examples are shown in Figs. 17 to 19 as an illustration. Figure 17 gives the absorption spectra of anthracene, naphthacene and pentacene. A constant planar geometry and symmetry (D_{2h}) is accompanied by an increase in the number of π-electrons which is noticable by a strong

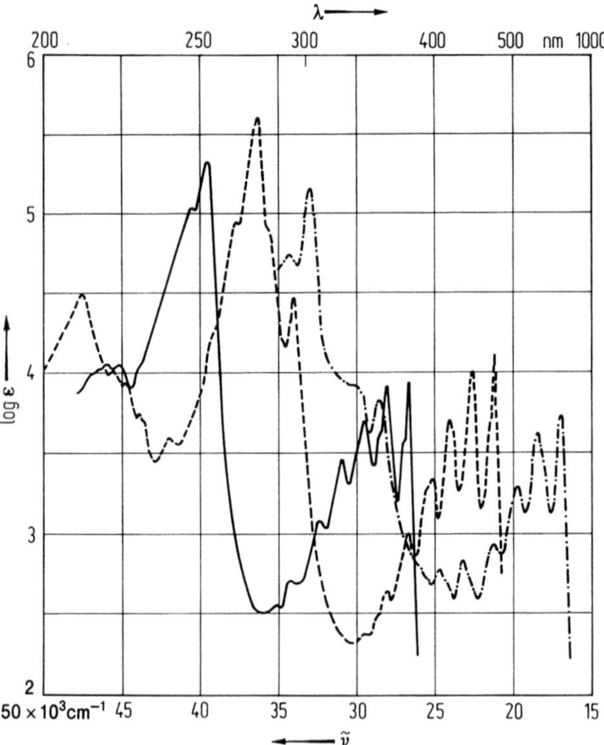

Fig. 17. Absorption spectra of anthracene (———), naphthacene (– – –), pentacene (–·–·–)

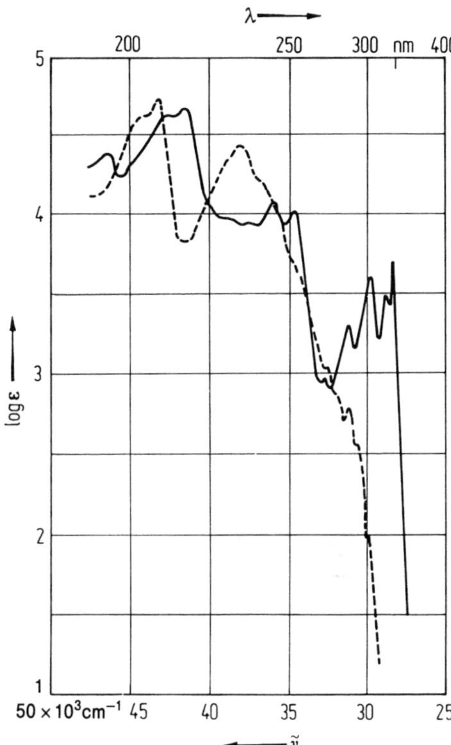

Fig. 18. Absorption spectra of 3,8-phenanthroline (———) and 1,10-phenanthroline (– – –)

displacement to the red, i.e. a reduced excitation energy. Figure 18 shows the absorption spectra of 3,8- and 1,10-phenanthroline. In both instances we have the same electron number and symmetry (C_{2v}) and only the geometry differs with regard to the position of the two N-atoms. By comparison with the absorption spectra of phenanthrene (Fig. 19) which has the same number of π-electrons and symmetry, the influence of the heteroatoms can be clearly recognized. Finally, if we look at the spectra of *trans*- and *cis*-stilbenes in Fig. 19 and at that of phenanthrene, we see how structure and symmetry clearly determine the absorption spectrum for the same number of π-electrons.

Pestemer [111], Timmons [172], Mataga and Kubota [173], Murrell [168], Suzuki [170], Fabian and Hartmann [174] and Sawicki [175] have all given detailed descriptions of the absorption spectra of organic compounds and of the influence of substituents on the spectra.

Notwithstanding the whole of computer technology, we are still restricted to collating spectra according to their similarity, i.e. according to specific principal structural characteristics. In accordance with this principle, 1200 example spectra have been compiled systematically in the DMS-UV-Atlas [176]. J. T. Clerc has published a more recent compilation of ca. 1000 UV-VIS spectra [177]. In order to permit a systematic classification by spec-

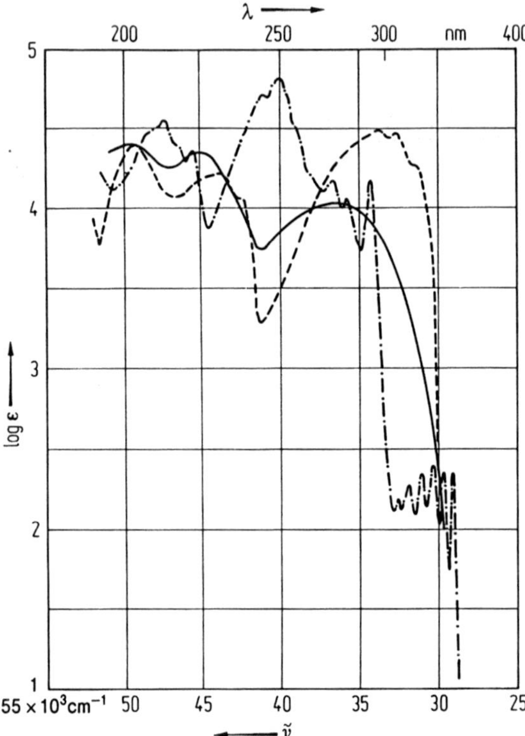

Fig. 19. Absorption spectra of *trans*-stilbene (– – –), *cis*-stilbene (———) and phenanthrene (–·–·–)

troscopic features, Pestemer et al. have developed a compound key in which they code each compound with a ten-digit number. This code then makes a systematic arrangement of spectra in a spectral file [178] possible.

With this key, each compound is characterized by a ten-digit figure. This corresponds to 10 groups of chromophoric and structural features which are arranged from left to right in order of decreasing "spectroscopic weight". Each of the ten digits contains three sub-criteria which permit a more accurate characterization by means of numbers 2, 3 and 4 in the digit place according to increasing spectroscopic weight. In the case of complicated compounds, these basic numbers can be combined as follows:

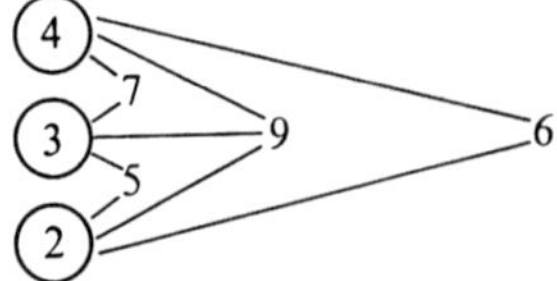

Thus, for each digit position there is a more accurate characterization with the numbers 0, 2, 3, 4, 5, 6, 7 and 9 where "0" always means: *no* spectroscopic feature at this digit. Numerals 1 and 8 are not used except at digit

3 where "8" denotes *free radicals*. Table 13 shows the classification of the BPPS key.

The sub-criteria at the first digit divide unsaturated compounds with UV-VIS active chromophores into unsaturated heterocycles (4), aromatic substances (3) and homocycles as well as C = C chains (2).

Digit 2 characterizes the type of condensation of ring systems. A "0" at this point means that the feature of the first digit 1 is not present in a con-

Table 13. Key for spectroscopically active structural elements of organic compounds

Digit place	Code number	Criteria	Examples
		Unsaturated rings and C–C multiple bonds:	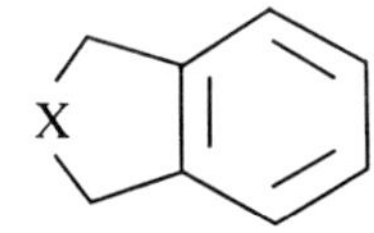 
(1)	4	Unsaturated heterocycles	
	3	Aromatic substances	
	2	Homocycles and chains with C=C bonds	
		Condensed aromatics and unsaturated heterocycles as well as homocycles	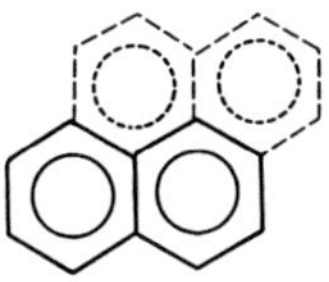 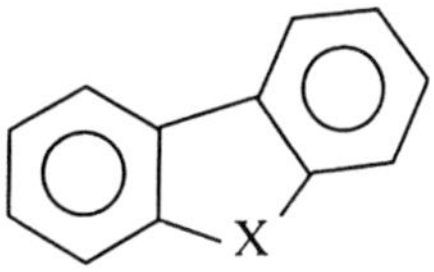
(2)	4	Angular or *peri*-condensed	
	3	Three or more rings, linearly condensed	
	2	Two rings	
		Conjugation and cumulative bonds	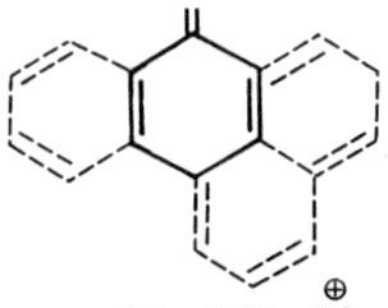 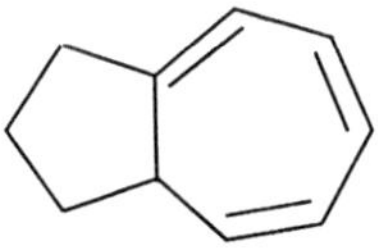
(3)	8	Free radicals	
	4	Cumulative bonds	
	3	Cross conjugation or *peri*-conjugation	
	2	Conjugation with charge resonance	$N-CH=\overset{\oplus}{N} \leftrightarrow \overset{\oplus}{N}=CH-N$
		Conjugation of:	
(4)	4	Aromatic rings (also heterocycles)	
	3	Triple bonds	
	2	Double bonds (in non-aromatic rings and chains) and three-membered rings (excluding heterocycles)	
		Conjugation with:	
(5)	4	Double-bonded heteroatoms in a conjugated system	–N=N–; –C=N–
	3	Double-bonded heteroatoms at the end of a conjugated system	–C=O; –C=NR; –C=S
	2	Single-bonded heteroatoms	{–C=C–X; X=NR_2, –OH, –OR, Halogen

Table 13 (continued)

Digit place	Code number	Criteria	Examples
		Heterocycles with:	
(6)	4	N (or other trivalent atoms)	
	3	S, Se, Te	
	2	O	
		Number of heteroatoms in a ring:	
(7)	4	Two or more different heteroatoms	
	3	Two or more of the same heteroatom	
	4	One heteroatom common to two rings	
		Ring Size (see Table 14)	
(8)	4	Three, four, seven or more atoms in a ring	
	3	Five atoms in a ring	
	2	Six atoms in a ring	
		Substituents	
(9)	4	multiple-bonded and containing double bonds	$-C=O$; $C=N-$; $-N=N$; SO_2
	3	singly bonded with double bonds	$-NO_2$; $-N=O$; $-C=N$
	2	containing only single bonds	$-NH_2$, $-NR_2$; $-OH$; $-OR$, Halogen, $-B$, Si
		Multiple occurrence of criteria (1) – (9)	
(10)	4	four or more times	
	3	three times	
	2	twice	

densed system. Thus, a simple division into condensed and non-condensed ring systems is possible:

Example:

Benzene 30.000/00.200
Naphthalene 32.000/00.600

Digits 3, 4 and 5 code the conjugation of the basic framework. If there is no conjugation then this is denoted with three times "0" (see examples of benzene and naphthalene). If the number 20000 appears in the first five digits then we are dealing with only *one* unconjugated double bond. The sequence 40000 denotes an isolated heterocyclic ring where we can then differentiate between oxygen (2), sulfur, selenium, tellurium (3) and nitrogen

(4) at digit 6. When using the key it has been shown that digits 1 and 6 must be used together for coding unsaturated heterocycles. The explanation of digit 6 in the original work concerning "saturated heterocycles" could easily lead to "0" being inserted in this position which actually allows a differentiation between the types of heteroatom.

Nonetheless, a differentiation between unsaturated and saturated heterocycles is easily achieved since the first five digits are occupied by a "0" in the case of a saturated heterocycle:

Pyridine 40.000/40.200
Piperidine 00.000/40.200

Digit 7 characterizes the heterocyclic ring system more accurately by number and type of heteroatom. The last three digits denote the ring size (8), substitution (9) and multiple occurrence of one of the preceding spectroscopic features (10). The following scheme is useful for feature (8):

Table 14. Examples of the determination of code-numbers at digit 8

Compound	Criteria	Code-No.
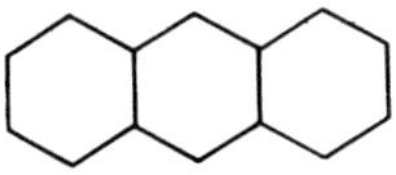	6- + 14-membered ring	2+4 = 6
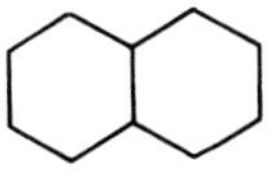	6- + 10-membered ring	2+4 = 6
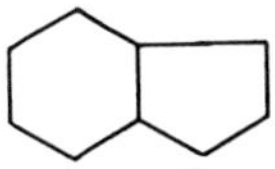	6- + 9-membered ring	2+3+4 = 9
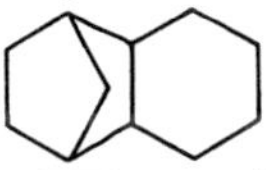	6- + 5- + 10-membered ring	2+3+4 = 9
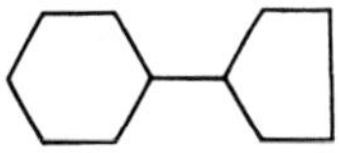	6- + 5-membered ring	2+3 = 5

Digit (10) can easily lead to discrepancies since it refers to the multiple occurrence of one feature at digits (9)→(1). If digit (9) = substituents is occupied then a multiple substitution should *always* be written at digit (10). If there is no substitution, unambiguity of the assignment in digit 10 can be achieved if it refers to the last digits (1)–(4) occupied by a number.

It is obvious that digits (5)–(8) are not typical for a multiple occurrence. If digit (10) is restricted to the first 4 digits a more accurate classification of the basic framework is attained:

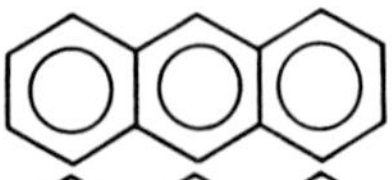	33.000/00.603	3 rings condensed
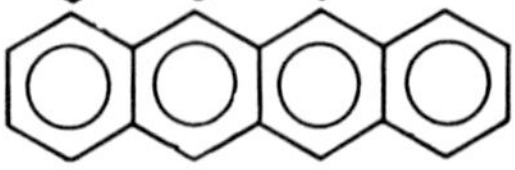	33.000/00.604	4 rings condensed
R–[CH = CH]$_n$–R		
n = 1	20.000/00.000	1 double bond
n = 2	20.020/00.002	2 double bonds
n = 3	20.020/00.003	3 double bonds
n = 4	20.020/00.004	4 double bonds
n = 5	20.020/00.005 etc.	5 double bonds
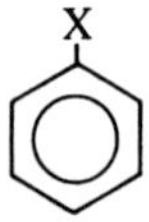	30.000/00.020	singly substituted
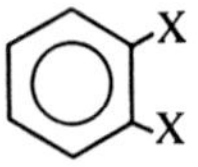	30.000/00.022	doubly substituted
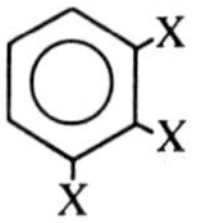	30.000/00.023 etc.	triply substituted

Further classification of the absorption spectra of organic compounds is possible by means of the *index number R* which gives the degree of unsaturation of an organic compound. We obtain R from the molecular formula by first calculating the total number of possible bonds. This is half the total number of valencies calculated as half the sum of the products of the number, n_i, of atoms of the i-th type and their corresponding valency, V_i,

$$\frac{1}{2} \sum n_i V_i \ .$$

We then substract the total number of bonds in an N-atom linear chain-type molecule, N-1 (this is the number, N, of all atoms less the first one):

$$R = \frac{1}{2} \sum n_i \cdot V_i - (N-1) \ . \tag{38}$$

We can illustrate this with the following examples:

1. *n*-hexane, C_6H_{14} $\quad R = \frac{1}{2} \times (6 \times 4 + 14 \times 1) - 19 = 0$,
2. Benzene, C_6H_6 $\quad R = \frac{1}{2} \times (6 \times 4 + 6 \times 1) - 11 = 4$,
3. Acridine, $C_{13}H_9N$ $\quad R = \frac{1}{2} \times (13 \times 4 + 9 \times 1 + 1 \times 3) - 22 = 10$,
4. Naphthaquinone-1,4, $C_{10}H_6O_2$ $\quad R = \frac{1}{2} \times (10 \times 4 + 6 \times 1 + 2 \times 2) - 17 = 8$.

If the compounds having the same R number are arranged according to their spectra, i.e. according to their associated λ_{max} or $\tilde{\nu}_{max}$ and ε_{max} values in a table, then, by looking up the tabulated values corresponding to measured λ_{max} and ε_{max} values, the question of the identity and structure of a compound can be quickly answered. A great quantity of information including the associated structural formulae, can be compiled in little space in such tables. Pestemer has drawn up such correlation tables [179] which are a valuable aid to structural analysis.

In addition to the lists of spectra mentioned there are other lists, some of which are continuously supplemented. However, in many cases they are not systematically compiled [180–187].

As far as inorganic compounds are concerned there are fewer detailed descriptions and collections than for organic compounds. However, the spectra of metal ions and their complexes, together with numerous inorganic or organic ligands, are of special importance and we are particularly interested in the two questions:

1. How is the spectrum of an organic ligand influenced by the central ion?
2. How is the spectrum of a central ion influenced by the ligand?

Question 1 is critically important in photometry involving chelating agents. Question 2 concerns the structure and symmetry of complexes directly and, thus, is of relevance for structure analysis. We can interpret the correlations theoretically by means of ligand-field theory. Schläfer and Gliemann [188, 189] and Lever [190] have presented detailed discussions of this theory and the interpretation of numerous spectra. The spectra of these complexes are important, not only in solution but also in the solid state, because their analysis reveals information about the structure of solid complexes.

4.4 Chemometrics

In recent years a group of mathematical methods such as those described in this chapter for processing UV-VIS spectral data has become known collectively as *chemometrics*. A rapid increase in interest in these chemometric methods has resulted in the publication of several texts [191–193] and the incorporation of appropriate computational packages into spectrometers supplied by all leading manufacturers. Each manufacturer has developed specific systems, usually based on methods described in the literature, and in addition there are independent data processing packages. The range of commercial systems for multicomponent analysis (Sect. 4.2) includes multilinear regression [194], maximum likelihood [194], partial least squares [196], iterative procedures and Kalman filter curve-fitting routines [197]. Fourier transform data processing has also been recommended [198]. Each of these routines has its advantages and limitations, but they all allow analysis of complex mixtures with an accuracy formerly impossible. Because these algorithms rely on distinguishing extremely small absorbance differences, critical attention must be paid to controlling any factor which can cause changes in spectra, such as pH and temperature variation and the quality of cuvette faces. The reliability of reference spectra is also important.

References

1. Lambert H (1852) Photometria, sive de mesura et gradibus luminis colorum et umbrae. 1760
2. Beer A (1852) Ann Physik 2 86:78
3. Pulfrich C (1925) Z Instrumentenkunde 45:116, 521
4. Löwe F (1949) Optische Messungen des Chemikers und des Mediziners. Techn Fortschrittsber, Bd 6. Steinkopff, Dresden, Leipzig
5. Kortüm G (1962) Kolorimetrie, Photometrie und Spektrometrie, 4. Auflg. Springer, Berlin Göttingen Heidelberg, Kap 1.5, S 21 ff
6. Thomas LC, Chamberlin GJ (1974) Calorimetric Chemical Analytical Methods, 8. Aufl. Tintometer Ltd. Salisbury, England. Wiley, London New York Toronto
7. Sutton D (1966–1971) In: Perkampus H-H, Sandemann I, Timmons CJ (Hrsg) DMS-UV-Atlas, Vol 5, Spektrum K1/14. Butterworth, London. Verlag Chemie, Weinheim, Vol I–V
8. Perkampus H-H, Kortüm K (1962) Z Analyt Chem 190:111
9. Ayres GH, Narang BD (1961) Anal Chim Acta 24:241
10. Sandell EB, Onishi H (1978) Photometric Determination of Traces of Metals, 3. Aufl. Interscience, New York
11. Iwantscheff G (1972) Chemical Analysis, Vol 3, Teil I. Das Dithizon und seine Anwendung in der Mikro- und Spurenanalyse, 2. Aufl. Verlag Chemie, Weinheim
12. Cheng KL, Bray RH (1955) Anal Chem 27:782
13. Berger W, Elvers H (1959) Z Anal Chem 171:185
14. Shibata S (1960) Anal Chim Acta 23:367, (1961) 25:348
15. Betteridge D, Fernando Q, Freiser H (1963) Anal Chem 35:294
16. Püschel R (1966) Z Anal Chem 221:132
17. Umland F, Hoffmann W (1959) Z Anal Chem 168:268
18. Umland F, Hoffmann W, Meckenstock KU (1960) Z Anal Chem 173:211
19. Umland F (1962) Anal Chem 190:186
20. Stary J (1962) Anal Chim Acta 28:132
21. Marczenko Z, Minczewski J (1960) Chem Anal (Warschau) 5:742; (1961) Rozniki Chem 35:1223; (1962) Zh Analit Khim 17:23
22. Marczenko Z, Kasinara K (1961) Chem Anal (Warschau) 6:37
23. Marczenko Z (1964) Bull Soc Chim France 939; (1964) Anal Chim Acta 31:224
24. Bartusek M, Okac A (1964) Collection Czech Chem Commun 26:52, 883, 2174
25. Becka J, Okac A (1971) ibid 36:2467, 3263
26. Smith GF (1954) 26:1534
27. Stephen WI (1969) Talanta 16:939
28. Schilt AA (1969) Analytical Applications of 1,10-Phenanthroline and Related Compounds. Pergamon, New York
29. Majumbar AK (1972) N-Benzoylphenylhydroxylamine and its Analogues. Pergamon, Oxford
30. Katyal M (1968) Talanta 15:95
31. Neoskaya ME, Nazarenko VA (1972) Zh Analit Khim 27:1699
32. Bock R (1951) Z Anal Chem 133:110
33. Rozycki C (1966) Chem Anal (Warschau) 11:447; (1970) 15:3
34. Rozycki C (1969) ibid 14:755

35. Rozycki C, Lachowicz E (1970) Chem Anal (Warschau) 15:255
36. Marczenko Z (1976) Spectrophotometric Determination of Elements. Wiley, New York London Sidney Toronto
37 Umland F (1971) Theorie und praktische Anwendung von Komplexbildnern. Akad Verlagsges. Frankfurt/M
38. Koch OG, Koch-Dedic GA (1974) Handbuch der Spurenanalyse, 2. Aufl, Teil 1 u 2. Springer, Berlin Heidelberg New York
39. Lange B, Vejdelek ZJ (1980) Photometrische Analyse. Verlag Chemie, Weinheim
40. Fries J, Getrost H (1975) Organische Reagenzien für die Spurenanalyse. E. Merck, Darmstadt
41. Delepine M. (1958) Bull Soc Chim France 5
42. Halls DJ (1969) Mikrochim Acta 62
43. Bode H (1954) Z Anal Chem 142:414; (1954) 143:182; (1955) 144:165
44. Hainberger L, Moreira SFT, Cotsas V (1977) Mikrochim Acta 303
45. Tsukahara I et al (1979) Anal Chim Acta 92:379
46. Uesugi K, Miyawaki M (1976) Mikrochim J 21:438
47. Buri BK, Burman S (1977) Fresenius Z Analyt Chem 286:253
48. Busev AI, Ivanov VM (1963) Z Anal Chim 18:208
49. Geary WJ et al (1962) Analyt Chim Acta 26:575
50. Stoner RE, Dasier W (1960) Analyt Chem 32:1207
51. Turkington RW, Tracy FM (1958) ibid 30:1699
52. Busev AI, Czan Fan (1961) Z Anal Chim 16:578
53. Shibata S (1960) Analyt Chim Acta 23:367
54. Cheng KL, Goydish BL (1963) Microchem J 7:166
55. Püschel R et al (1966) Z Analyt Chem 223:44
56. Rinehart RW (1954) Analyt Chem 26:1820
57. Young JP et al (1960) Analyt Chem 32:928
58. Budesinsky B, Haas K (1965) Z Analyt Chem 210:263
59. Young JP et al (1958) Analyt Chem 30:422
60. Flaschka H, Farah MY (1956) Z Analyt Chem 152:401
61. Kahrt L et al (1979) Chem Abstr 90:15790q
62. Shtokalo MI et al (1978) Chem Abstr 89:190443e
63. Savvin SB (1961) Talanta 8:673
64. Luke CL (1953) Analyt Chem 25:674
65. McNulty BJ, Wollard LD (1955) Analyt Chim Acta 13:64
66. Onishi H, Sandell EB (1954) ibid 11:444
67. Ramette RW, Sandell EB (1955) ibid 13:455
68. White CE, Rose HJ (1953) Analyt Chem 25:351
69. Polucktov NS et al (1958) Z Anal Chim 13:396
70. Ducret L (1959) Analyt Chim Acta 21:86
71. Babko AK et al (1966) Z Anal Chim 21:196
72. Ducret L (1957) Analyt Chim Acta 17:213
73. Pasztor L, Bode JD (1960) Analyt Chem 32:1530
74. Marczenko Z (1977) Mikrochim Acta (Wien) II 651
75. Marczenko Z, Kowalczyk E (1979) Analyt Chim Acta 108:261
76. Marczenko Z, Uscinska J (1981) ibid 123:271
77. Marczenko Z, Jarosz M (1981) Analyst 106:751
78. Aoyama M et al (1981) Analyt Chim Acta 129:237
79. Svoboda N, Chromy V (1966) Talanta 13:237
80. Thierig D, Umland F (1965) Z Analyt Chem 211:161
81. Umland F et al (1960) Z Analyt Chem 173:211
82. Umland F, Janssen A (1970) Z Analyt Chem 249:186
83. Umland F et al (1966) ibid 215:401
84. Tôei K et al (1981) Analyst 106:776
85. Hora FB, Webber PJ (1960) Analyst 85:567
86. Hartley AM, Asai RI (1963) Analyt Chem 35:1207

87. Barnes H (1950) Analyst 75:388
88. Nelson JL et al (1950) Anal Chem 26:1081
89. Morris AW, Riley JP (1963) Anal Chim Acta 29:272
90. Roth H (1951) Mikrochem/Mikrochim Acta 36/37:379
91. Johnson CM, Nishita H (1952) Analyt Chem 24:736
92. Gustaffson L (1960) Talanta 4:227
93. Sinclair A et al (1971) Talanta 18:972
94. Nakamura M, Murata A (1979) Analyst 104:985
95. Nakamura M (1981) ibid 106:493
96. Baca P, Freise H (1977) Analyt Chem 49:2249
97. Flamerz S, Bashir WA (1981) Analyst 106:243
98. Dtsch. Einheitsverf. z. Wasser-, Abwasser- u. Schlammuntersuchung; J. Lief. (1975) Verlag Chemie J. Lief., Weinheim
99. Richtlinien zur Aufstellung von Probennahmeprogrammen, Normenausschuß Wasserwesen (1980) Internat Norm 150, 5667-1. Dtsch Inst f Normung. Beuth, Berlin
100. Soulard M et al (1981) Analysis 9:35
101. Capelle R (1961) Anal Chim Acta 24:555; (1961) 25:59
102. Freier RK (1974) Wasseranalysen, 2. Aufl. de Gruyter, Berlin New York
103. Hein H (1978) Angew. UV-Spektroskopie, Heft 5. Bodenseewerk Perkin-Elmer
104. Franke G, Hein H (1979) Chemie-Technik 185, 295
105. Analysen von Oberflächen- und Abwasser mit dem Filter-Photometer Nanocolor 25. Macherey-Nagel & Co, D-5160 Düren
106. Roberts JD, Green C (1946) J Am Chem Soc 68:214
107. Wells CF (1966) Tetrahedron 22:2685
108. Brande EA (1945) J Chem Soc 498
109. Roberts JD, Green C (1946) J Am Chem Soc 68:214
110. Bohlmann F (1951) Chem Ber 84:490
111. Pestemer M (1955) In: Houben-Weyl, Methoden der Organischen Chemie Bd III, 2, 4. Aufl. Thieme, Stuttgart
112. Then R, Radler F (1968) Z Lebensmittelunters Forsch 138:163
113. Heintze K, Braun F (1970) ibid 142:40
114. Sawicky E et al (1961) Mikrochem J 5:225
115. Sawicky E et al (1967) Anal Clin Acta 39:505
116. Sawicky E et al (1961) Anal Chem 33:93
117. Sawicky E, Sawicky CR (1975) Aldehydes-Photometric Analyse, Vol 1 and 2; Vol 5 (1978). Academic Press, London New York San Francisco
118. Kakac B, Vejdelek DJ (1974) Handbuch der photometrischen Analyse organischer Verbindungen, Bd 1; Bd 2 (1974); Ergänzungsband (1977). Verlag Chemie, Weinheim
119. Merck E (Hrsg) (1974) Klinisches Labor: Medizinisch-chemische Untersuchungsmethoden, 12. Aufl. E. Merck, Darmstadt
120. Zeiss C (Hrsg) (1967–1971) Photometrische Mikromethoden, Klinische Chemie. Carl Zeiss, Oberkochem
121. Boehringer (Hrsg) (1980) Testfibel. Boehringer Mannheim GmbH, Diagnostica
122. Leithe W (1972) Die Analyse organischer Verbindungen in Trink-, Brauch- und Abwässern. Wiss Verlagsges, Stuttgart
123. Merck E (Hrsg) (1975) Die Untersuchung von Wasser. Darmstadt
124. Pesez M, Bartos J (1975) Colorimetric and Fluorimetric Analysis of organic Compounds and Drugs. M. Dekker, New York
125. Gstirner F (1951) Chemisch-physikalische Vitaminbestimmungsmethoden, 4. Aufl. Enke, Stuttgart
126. Hashmi MH (1973) Assays of Vitamin in Pharmaceutical Preparation. Wiley, London New York Sydney Toronto
127. Knobloch E (1963) Physikalisch-chemische Vitaminbestimmungsmethoden. Akademie, Berlin
128. Knorr F (1961) Vitaminbestimmungsmethoden. Moser, Garmisch-Partenkirchen
129. Merck E (Hrsg) (1960) Vitaminanalyse. E. Merck, Darmstadt

130. Bartos J, Pesez M (1977) Colorimetric and Fluorimetric Analysis of Steroids. Academic Press, London
131. Bergmeyer HU (1962) Methoden der enzymatischen Analyse. Verlag Chemie, Weinheim
132. Long EC: Varian Instruments Appl. Report, CPT-2223
133. Boehringer (Hrsg) (1981) Arbeitsanleitung, Makro-Technik. Boehringer Mannheim GmbH, Diagnostica; Boehringer (Hrsg) (1981) Arbeitsanleitungen, Mikro-Technik. Boehringer Mannheim GmbH, Diagnostica
134. Mattenheimer H (1976) Die Theorie des enzymatischen Tests. Boehringer Mannheim GmbH, Diagnostica
135. Cornish-Bowden A (1979) Fundamentals on Enzyme Kinetics. Butterworth, London
136. Bergmeyer HU, Gawehn K (Hrsg) (1977) Grundlagen der enzymatischen Analyse. Verlag Chemie, Weinheim New York
137. Briggs GE, Haldane JBS (1925) Biochem J 19:338; s. auch Hammett LP: Physikalische organische Chemie. Verlag Chemie, Weinheim, S 83–84
138. Michaelis L, Menten ML (1913) Biochem Z 49:333
139. Chance BJ (1940) Franklin Inst 228:459
140. Bautler E et al (1980) Varian UV-VIS-Spectrophotometry, Nr UV-5, Aug 1980
141. Moody GW, Heisz O (1980) Labor-Praxis in der Medizin 4, Heft 4
142. Boehringer (Hrsg) (1980) Methoden der enzymatischen Lebensmittelanalytik. Boehringer Mannheim GmbH, Biochemica
143. Brummer WN, Ebeling N (1976) Merck-Kontakte 2/76, S 3–7
144. Henniger G (1979) Z Lebensm-Technol u Verfahrenstechnik 30:137
145. Krüger E, Nordmann A (1981) Labor-Praxis in der Medizin 5, Heft 1–2
146. Molch D et al (1975) Z Chem 15:410; (1976) 16:109
147. Harker III GG et al (1970) J Chem Educ 47:712
148. Sternberg JC et al (1960) Anal Chem 32:84
149. Baummann RP (1962) Absorption Spectroscopy. Wiley, New York London, Kap 9.4, S 403 ff
150. Zscheile FP et al (1962) Anal Chem 34:1776
151. Herschberg IS (1964) Fresenius Z Anal Chem 205:180
152. Herschberg IS, Sicma FLJ (1962) Koninkl Ned Akad Wetenschap, Proc Ser B 65:244, 256
153. Cerfontain H et al (1963) Anal Chem 35:1005
154. Arends JM et al (1964) Anal Chem 36:1802
155. Doerffel K et al (1966) Z Chem 6:155
156. Sustek J (1974) Analyt Chem 46:1676
157. Leggett DJ, McBryde WAE (1975) Analyt Chem 47:1065
158. Leggett DJ (1977) Analyt Chem 49:276
159. Fahr E, Schmid M (1980) Fresenius Z Anal Chem 300:381
160. James GF (1980) Techn Paper UV-1. Hewlett Packard Co
161. Long EC, Greig D (1980) Varian UV-VIS Spectrophotometry Nr UV-9
162. Jochum C et al (1981) Anal Chem 53:85
163. Aldous KM et al (1975) ibid 47:1034
164. Kontron-Analytik-Schrift (1980) Photometrie – Basis Information
165. Dewer MJS, Urch DS (1975) J Chem Soc 345
166. Bergmann G (1981) Mehrkomponentenanalyse, Plenarvortrag Arbeitstagung prakt. Molekülspektroskopie, Dortmund
167. Jaffé H-H, Orchin M (1962) Theory and Applications of Ultraviolet Spectroscopy. Wiley, New York London
168. Murrell JN (1963) The Theory of the Electronic Spectra of Organic Molecules. Methuen, London
169. Orchin M, Jaffé H-H (1971) Symmetry, Orbitals and Spectra, Supplement. Wiley, New York London Sydney Toronto
170. Suzuki H (1967) Electronic Absorption Spectra and Geometry of Organic Molecules. Academic Press, New York London
171. Platt JR (1964) Systematics of the Electronic Spectra of Conjugated Molecules. Wiley, New York

172. Timmons CJ, Stern ES (1970) Gillam and Stern's Introduction to Electronic Absorption Spectroscopy in Organic Chemistry, 3. Aufl. Edwards Publ, London
173. Mataga N, Kubota T (1970) Molecular Interactions and Electronic Spectra. Dekker, New York
174. Fabian J, Hartmann H (1980) Light Absorption of Organic Colorants. Springer, Berlin Heidelberg New York
175. Sawicki E (1970) Photometric Organic Analysis. Wiley, New York London Sydney Toronto
176. Perkampus H-H, Sandemann I, Timmons CJ (Hrsg) DMS-UV-Atlas. Butterworth, London. Verlag Chemie, Weinheim, Vol I–V, 1966–1971
177. Clerc JT (1983) UV-Spektrensammlung, on Discettes. Verlag Chemie, Weinheim
178. Bergmann G, Pestemer M, Perkampus H-H, Schreder B. Angew Chemie Int Ed
179. Pestemer M (1974) Correlation Tables for the Structural Determination of Organic Compounds by Ultraviolet Light Absorptiometry. Verlag Chemie, Weinheim
180. Organic Electronic Spectral Data. Interscience, New York 1960–1969
181. Lang L (ed) Absorption Spectra in UV and Visible Region. Acad Press, New York, Vol 1–5, 1961–1965; Akadémiai Kiadó, Budapest, Vol 6–13, 1965–1969
182. Neudert W, Röpke H (1965) Atlas of Steroid Spectra. Springer, Berlin Heidelberg New York
183. Catalog of Ultraviolet Spectral Data Manufacturing Chemist'. Ass Res Proj Chem Thermodynamic Properties Center, Texas, 1964ff
184. Handbook of Ultraviolet and Visible Absorption Spectra of Organic Compounds (Hirayama) (1967) Plenum Press, New York
185. Sadtler Spectra; Sadtler Research Laboratories, Philadelphia
186. Friedel RA, Orchin M (1951) Ultraviolet Spectra of Aromatic Compounds. Wiley, New York
187. Ultraviolett Spectraldata; Amer Petroleum Inst Res Proj 44, Carnegie Inst and US Bureau of Stand
188. Schläfer HL, Gliemann G (1967) Einführung in die Ligandenfeldtheorie. Akad Verlagsges, Frankfurt/M
189. Schläfer HL, Gliemann G (1969) Basic principles of Ligand field theory. Wiley, New York
190. Lever ABP (1984) Inorganic electronic spectroscopy, 2nd edn. Elsevier, New York
191. Martens H, Naes T (1989) Multivariate Calibration. Wiley, New York
192. Massart DL, Vandeginste BGM, Denning SN, Michotte Y, Kaufman L (1988) Chemometrics – A Textbook. Elsevier, Amsterdam
193. Brereton RG (1990) Chemometrics. Ellis Horwood, Chichester
194. Brown CW, Lynch PF, Obremsko RJ, Lavery DS (1982) Anal Chem 54:1472
Fredericks PM, Lee JB, Osborn PR, Swinkel DAJ (1985) Appl Spectr 39:303
195. Beck JV (1977) Parameter Estimation in Engineering and Science. Wiley, New York
196. Haarland DM, Thomas EV (1988) Anal Chem 60:1193
Wold S, Ebersen K, Geladi P (1987) Chemometrics and Intelligent Laboratory Systems 2:37
197. Tranter RS (1990) Anal Proc 27:26
198. Gilbert AS (1987) In: Burgess C, Mielenz KD (eds) Advances in standards and methodology in spectrophotometry. Elsevier, Amsterdam

5 Recent Developments in UV-VIS Spectroscopy

In recent years we have seen the development of many new techniques such as:

Dual-wavelength spectroscopy,
derivative spectroscopy,
reflection spectroscopy,
photoacoustic spectroscopy,
luminescence-excitation spectroscopy,
variable wavelength detectors for chromatography
and enzyme kinetics

all of which quickly awakened great interest.

The basics of enzyme kinetics have been detailed in Sect. 4.1.5.

In the following paragraphs, the principles of dual-wavelength, derivative, reflection, photoacoustic and luminescence-excitation spectroscopy are described briefly and some examples of their applications are given.

5.1 Dual-Wavelength Spectroscopy

Dual-wavelength spectroscopy, also called two-wavelength spectroscopy, is extremely versatile [1]. We are dealing here with wavelengths λ_1 and λ_2 and we can determine an absorbance difference related to λ_1 at λ_2.

$$\Delta A = A_{\lambda_2} - A_{\lambda_1} \ .$$

In analysis it is an important fact that an absorbance difference, ΔA, in a binary mixture of components a and b is proportional to the concentration of only one component if we have selected the correct wavelength λ_1. We can prove this without difficulty.

In a two-component system with concentrations c_a and c_b, the following equations apply to the absorbance at both wavelengths, λ_1 and λ_2 (d = 1 cm):

$$\begin{aligned} A_1 &= \varepsilon_{1a} c_a + \varepsilon_{1b} c_b = A_{1,a} + A_{1,b} \ , \\ A_2 &= \varepsilon_{2a} c_a + \varepsilon_{2b} c_b = A_{2,a} + A_{2,b} \ . \end{aligned} \tag{39}$$

ε_{1a} and ε_{1b} are the molar decadic extinction coefficients of components a and b at wavelength λ_1; correspondingly, ε_{2a} and ε_{2b} at wavelength λ_2.

In a dual-wavelength measurement the absorbance difference is obtained:

$$A_2 - A_1 = \Delta A = (\varepsilon_{2a} - \varepsilon_{1a})c_a + (\varepsilon_{2b} - \varepsilon_{1b})c_b \ . \tag{40}$$

If component b has the same extinction coefficient ε at both wavelengths λ_1 and λ_2; the second term in brackets in Eq. (40) equals zero:

$$\varepsilon_{2b} \cdot c_b - \varepsilon_{1b} \cdot c_b = A_{2,b} - A_{1,b} = 0 \tag{41}$$

and we obtain

$$\Delta A = (\varepsilon_{2a} - \varepsilon_{1a})c_a \ . \tag{42}$$

Thus, the absorbance difference ΔA is only a function of component a and the influence of component b can be eliminated! If component a does *not* absorb at wavelength λ_1, then $\varepsilon_{1a} = 0$ and Eq. (42) can be written in the simplified form:

$$\Delta A = \varepsilon_{2a} \cdot c_a = A_{2,a} \ . \tag{42a}$$

Equations (42) and (42a) form the basis of the *equiabsorption* or *equiextinction* method.

The wavelengths λ_1 and λ_2 can be selected from the known absorption spectra of components a and b. Figure 20 illustrates the application of this technique to the absorption spectra of components a and b and their superposition in a mixture. The following evaluation is based on Eq. (42a).

An example of the application of Eq. (42a) is shown in Fig. 21, i.e. the binary system $NaNO_3$ and $NaNO_2$ in water. The NO_3^- ion shows an absorption maximum at 302 nm ($\tilde{\nu} = 33112 \text{ cm}^{-1}$) with $\varepsilon_{NO_3^-} = 7.0 \text{ l mol}^{-1} \text{ cm}^{-1}$.

The NO_2^- ion has an extinction coefficient of $\varepsilon_{NO_2^-} = 9 \text{ l mol}^{-1} \text{ cm}^{-1}$ at the same wavelength [2]. In addition, the NO_2^- ion has a long wavelength band with a maximum at $\lambda = 355$ nm (28169 cm^{-1}). The NO_3^- ion does not absorb within this band and we can conveniently determine reference wavelength λ_1 by means of Fig. 21.

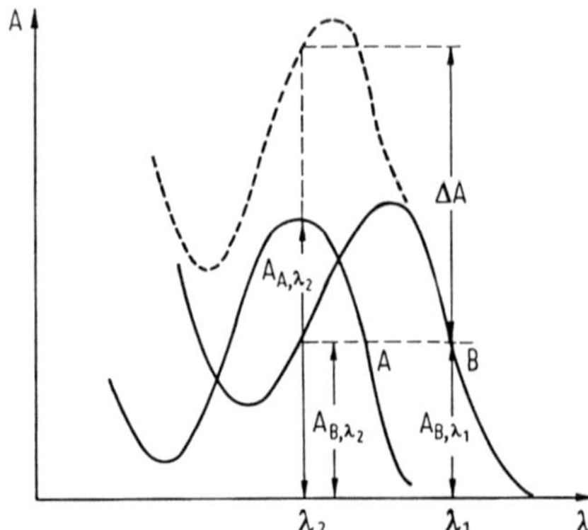

Fig. 20. Schematic diagram of the dual-wavelength equiextinction method (see text for explanation)

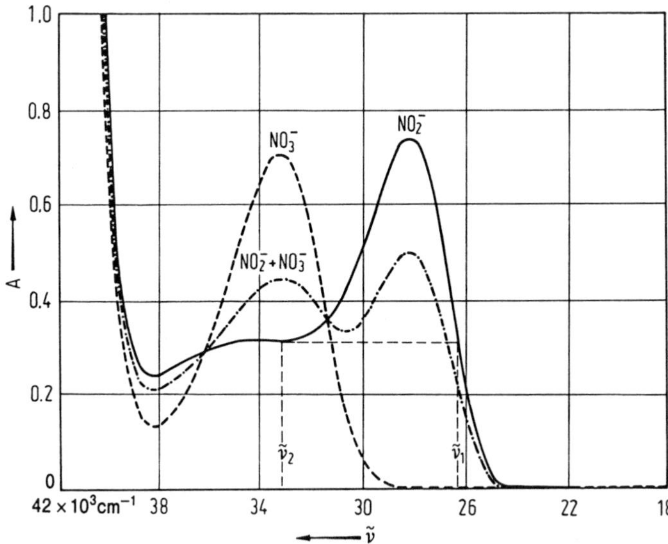

Fig. 21. Example of the selection of analytical wavelengths in dual-wavelength spectroscopy; NO_3^-/NO_2^- system in H_2O; NO_2^- (———), NO_3^- (– – –), mixture (–.–.–)

Absorbance $A_{1,b}$ then corresponds exactly to absorbance $A_{2,b}$ and the absorbance difference at $\tilde{\nu}_2$ determined according to Eq. (42a) is directly proportional to the concentration of the NO_3^- ions.

Figure 22 shows the spectra of different NO_3^-/NO_2^- mixtures where the concentration of the NO_2^- ion was kept constant. The dual-wavelength positions λ_1 and λ_2 ($\tilde{\nu}_1$ and $\tilde{\nu}_2$) can be seen directly in this figure. Figure 23 shows the linear correlation of the measurement in Fig. 22 calculated using Eq. (42a). This example illustrates the analytical problem of a two-component system, i.e. the quantitative determination of NO_3^- in the presence of NO_2^-.

The *hydroquinone* and *phenol* system is another example of the application of the equiextinction or equiabsorption method [3]. Schmitt has also described the hydroquinone/pyrocatechol, hydroquinone/resorcinol, vitamin B6/nicotinamide and phenylalanine/tryptophan systems [3]. The prerequisite is that the absorption maximum of component a (e.g. NO_3^-) overlaps the absorption flank of the accompanying component b (e.g. NO_2^-) whilst component a does not underly component b (see Figs. 21 and 22). Then component b can be determined by direct spectrophotometric means. However, the measurement of component a would result in a considerable positive error because a contribution from the absorption of b is also measured. By the equiextinction method, this interfering influence of component b can be eliminated with the help of dual-wavelength spectroscopy. With this method, we measure with only one cuvette, i.e. no reference solution is required. The greatest advantage lies in the fact that

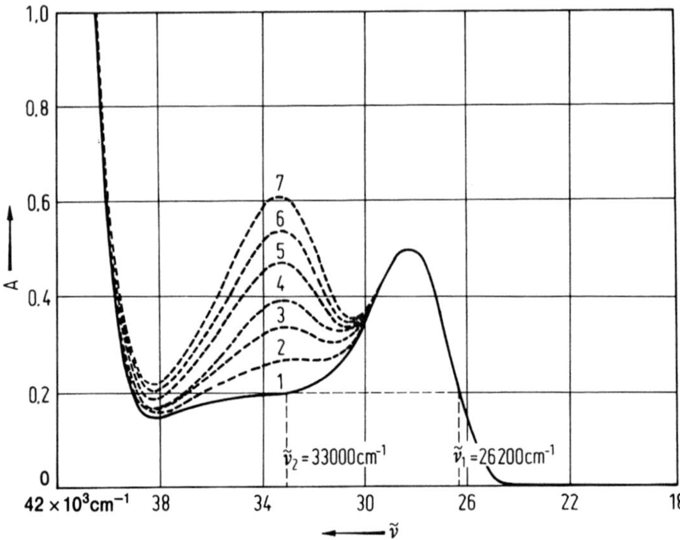

Fig. 22. Absorption spectra of mixtures of the NO_3^-/NO_2^- system; $C_{NO_2^-}$: constant; $C_{NO_3^-}$: variable (curves 2–7)

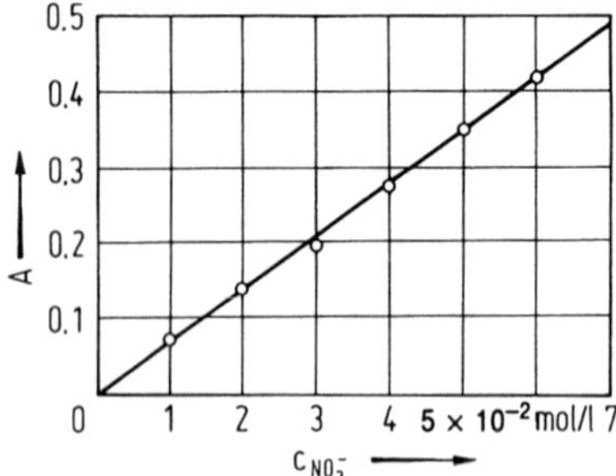

Fig. 23. Calibration line for Fig. 22

background interferences due to turbidity and/or changes with time can be eliminated. This is very difficult to achieve with standard absorption spectroscopy.

If turbidity is present, i.e. there are scattering particles in the solution, we measure not only the true absorbance, A_w, but also an apparent absorbance, A_s, i.e. a strong background results as shown in simplified form in Fig. 24.

It can be seen immediately that Eq. (40a) may be obtained by analogy with (40):

$$\Delta A = (A_{s,\lambda_2} + A_{w,\lambda_2}) - A_{\lambda_1} = A_{w,\lambda_2} , \quad (40a)$$

since according to Fig. 24, $A_{s,\lambda_2} - A_{\lambda_1} = 0$.

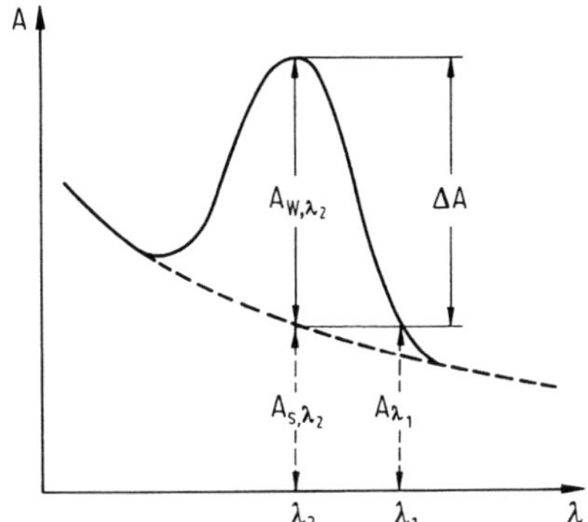

Fig. 24. Dual-wavelength spectroscopy in the presence of a scattering background

Equation (40a) shows that the apparent absorption at wavelength λ_2 is determined by measuring at wavelength λ_1. Thus, A corresponds to the true absorbance at wavelength λ_2.

Figures 20 to 24 show immediately the significance of this method in analysis. Chance [4, 5] first developed dual-wavelength spectroscopy specifically in order to analyze *turbid solutions* of biological and physiological systems quantitatively. Shibata has discussed its general application in analytical chemical problems [1]. Since the sample and reference cell are identical the smallest absorbance can be determined reliably in a full scale deflection range of $0.001 \leq A \leq 0.010$. The quantitative determination or detection of substances can be extended to the ppb region which is particularly valuable in environmental analysis (the limit of detection for mercury with dithizone in aqueous solution is 1.8 ppb [6]).

The absorption of a *complexing agent* itself can be utilized in quantitative photometric determinations with complexing agents. The reference wavelength λ_1 is set to the maximum of the absorption spectrum of the complexing agent and λ_2 to the maximum of the absorption spectrum of the complex. As a result of a complex formation the absorbance of the complexing agent at wavelength λ_1 decreases, i.e. we obtain a negative absorbance whilst the absorbance of the complex at wavelength λ_2 increases. Both effects are proportional to the concentration and the absorbance difference ΔA is given by

$$\Delta A = A_{\lambda_2} - (-A_{\lambda_1}) = A_{\lambda_2} + A_{\lambda_1} \ . \quad (43)$$

This produces an apparent increase of the absorbance of the complex which leads to an increased sensitivity of detection. Shibita et al. have described this method using a cobalt determination with 4-[(3,5-dichlor-2-pyridyl)azo]-*m*-phenylene-diamine as an example [7, 8].

Chemical reactions in turbid solutions can also be followed by this method. Here we select λ_1 as the reference wavelength at one of the isosbestic points. However, in the case of turbid solutions, the measuring wavelength λ_2 should lie in the proximity of λ_1 because we know that scattering in

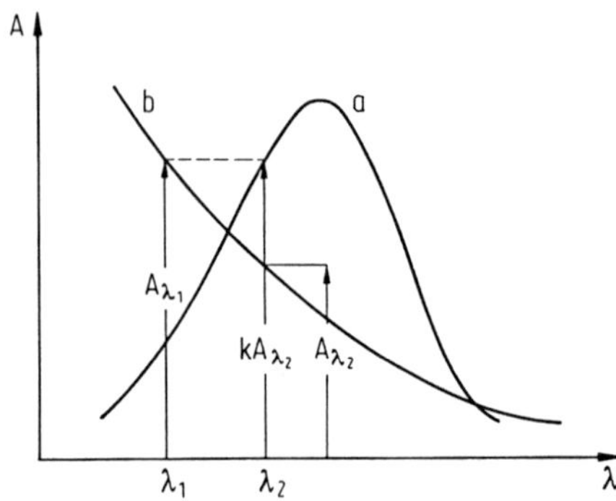

Fig. 25. Overlaying of absorption spectra and the signal-amplification method

such solutions is a function of the wavelength. The dependence of scattering upon wavelength allows measurement of the absorption spectrum of a turbid solution only in simple cases. By reference against the fixed wavelength λ_1 with scattering plus absorbance A_{s,λ_1} a relative absorption spectrum can be obtained by scanning λ_2.

If the above requirements are not met, i.e. extinction coefficients ε_{1b} and ε_{2b} differ at every combination of wavelengths λ_1 and λ_2 we simply modify the signal, thereby making it possible for the absorbance difference to become independent of the concentration of component b. Figure 25 shows an example of this. The uniformity of absorbances A_{λ_1} and A_{λ_2} assumed for the equiextinction method is achieved by instrumental means. Most UV-VIS spectrophotometers fitted with microprocessors or microcomputers permit such conversions without difficulty.

$\varepsilon_{1b} \cdot c_b = A_{1b}$ applies at wavelength λ_1 but

$\varepsilon_{2b} \cdot c_b = A_{2b}$ applies at wavelength λ_2 .

It can be seen immediately from Fig. 25 that $A_{1b} \neq A_{2b}$.

If A_{2b} is multiplied by factor k the following applies

$$\varepsilon_{1b} c_b = k \cdot \varepsilon_{2b} c_b \ . \tag{44}$$

Thus, in principle the assumption of the equiextinction method has again been met.

Now, the following applies to the measured extinction difference:

$$A_{\lambda_2} - A_{\lambda_1} = k(\varepsilon_{2b} c_b + \varepsilon_{2a} c_a) - (\varepsilon_{1b} c_b + \varepsilon_{1a} c_a) \ , \tag{45}$$

$$A_{\lambda_2} - A_{\lambda_1} = (k\varepsilon_{2b} - \varepsilon_{1b}) c_b + (k\varepsilon_{2a} - \varepsilon_{1a}) c_a \ . \tag{45a}$$

And using the condition (44) we have

$$A_{\lambda_2} - A_{\lambda_1} = \Delta A = (k\varepsilon_{2a} - \varepsilon_{1a}) c_a \ . \tag{46}$$

The absorption difference ΔA in Eq. (46) is again independent of the concentration of component b. Therefore, component a can be determined quantitatively in a mixture.

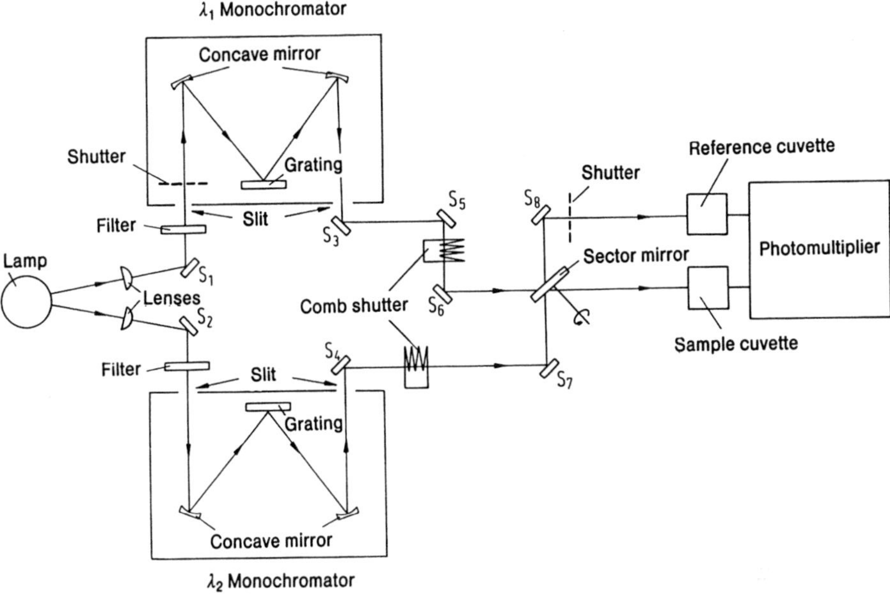

Fig. 26. Optical system of the Hitachi-Perkin-Elmer dual-wavelength spectrophotometer, model 557

In practice, factor k in Eq. (44) is equivalent to an amplification of the signal at wavelength λ_2 and we call this technique the *signal-amplification method*. It has also been developed for three-component systems by Honkawa [9].

Any standard single-beam photometer can be used for the equiextinction method shown in Figs. 21 to 23. The absorbances A_{λ_2} and A_{λ_1} are measured one after the other and A_{λ_1} is deducted as a reference value from A_{λ_2} [3].

Typical dual-wavelength spectrometers are fitted with two monochromators as shown by the outline of the light path in Fig. 26. The diagram here is a simplified reproduction of the light path of the Hitachi-Perkin-Elmer dual-wavelength spectrophotometer, model 557. This instrument can also be used as a conventional dual-beam spectrophotometer with a sample and reference cuvette.

In this case, the shutter of the λ_1 monochromator is closed and the one after the angle mirror S_8 is opened. The sector mirror ensures that the monochromatic beam of light from the λ_2 monochromator passes first through the sample cuvette and then through the reference cuvette. When used as a dual-wavelength instrument, both beams of light coming from the deuterium (UV) lamp or tungsten-iodine (VIS) lamp pass through both monochromators if the shutter of the λ_1 monochromator is open.

If the shutter behind S_8 is closed, the light from the λ_2 monochromator can only fall onto the sample cuvette via the sector mirror whilst the path of light from the λ_1 monochromator also falls onto the sample cuvette if the sector mirror is rotated by 180°. Thus, both beams of light fall on the photomultiplier with a periodicity determined by the rotation frequency of the sector mirror. They generate an alternating signal in the sequence λ_1 signal, zero signal, λ_2 signal, zero signal, λ_1 signal ... which goes, via a chopper synchronized with reference signals, to the measurement channels provided for the individual signals. The process is controlled by microprocessors. Modern electronic measurement technology ensures a reduction of the noise by a factor of 2×10^4.

5.2 Derivative Spectroscopy

Derivative spectroscopy is the representation of the first and second derivative as well as higher derivatives of an absorbance with respect to the wavelength as a function of the wavelength, i.e.

$$dA/d\lambda,\ d^2A/d\lambda^2,\ d^3A/d\lambda^3 \ldots \quad \text{as } f(\lambda)\ .$$

Between 1953 and 1955 Hammond et al. [10], Morrison [11] and Giese and French [12] introduced this method, which is becoming increasingly important.

Differentiating the Bouguer-Lambert-Beer law as shown in Eq. (4) the following equations are obtained:

zero order derivative:

$$A = \varepsilon \times c \times d\ , \tag{4}$$

first derivative:

$$\frac{dA}{d\lambda} = c \cdot d \cdot \frac{d\varepsilon}{d\lambda}\ , \tag{47}$$

second derivative:

$$\frac{d^2A}{d\lambda^2} = c \cdot d \cdot \frac{d^2\varepsilon}{d\lambda^n}\ , \tag{47a}$$

nth derivative:

$$\frac{d^nA}{d\lambda^n} = c \cdot d \cdot \frac{d^n\varepsilon}{d\lambda^n}\ . \tag{47b}$$

Equations (47) to (47b) show immediately that derivatives $d^nA/d\lambda^n$ are always proportional to the concentration and analytical applications are based on this fact.

In contrast, if we form derivatives by using the Bouguer-Lambert-Beer law in the antilogarithmic form with ε_n as the natural molar extinction coefficient

$$I = I_0 e^{-\varepsilon_n \cdot c \cdot d} \tag{48}$$

we obtain

first derivative:

$$\frac{1}{I_0}\left(\frac{dI}{d\lambda}\right) = -c \cdot d \cdot \frac{d\varepsilon_n}{d\lambda} \cdot e^{-\varepsilon_n c \cdot d} \tag{48a}$$

or with (48)

$$\frac{1}{I}\left(\frac{dI}{d\lambda}\right) = \frac{d \ln I}{d\lambda} = -c \cdot d \cdot \frac{d\varepsilon_n}{d\lambda}\,, \tag{48b}$$

second derivative:

$$\frac{1}{I_0}\left(\frac{d^2 I}{d\lambda^2}\right) = c^2 d^2 \left(\frac{d\varepsilon_n}{d\lambda}\right)^2 e^{-\varepsilon_n c \cdot d} - c \cdot d \left(\frac{d^2 \varepsilon_n}{d\lambda^2}\right) e^{-\varepsilon_n c \cdot d} \tag{49}$$

or

$$\frac{1}{I}\left(\frac{d^2 I}{d\lambda^3}\right) = -c \cdot d \left(\frac{d^2 \varepsilon_n}{d\lambda^2}\right) + c^2 d^2 \left(\frac{d\varepsilon_n}{d\lambda}\right)^2, \tag{49a}$$

third derivative:

$$\frac{1}{I_0}\left(\frac{d^3 I}{d\lambda^3}\right) = \left[-c \cdot d \left(\frac{d^3 \varepsilon_n}{d\lambda^3}\right) + 3c^2 d^2 \left(\frac{d\varepsilon_n}{d\lambda}\right) \cdot \left(\frac{d^2 \varepsilon_n}{d\lambda^2}\right) - c^3 d^3 \left(\frac{d\varepsilon_n}{d\lambda}\right)^3\right] e^{-\varepsilon_n c \cdot d}$$

or

$$\frac{1}{I}\left(\frac{d^3 I}{d\lambda^3}\right) = -c \cdot d \left(\frac{d^3 \varepsilon_n}{d\lambda^3}\right) + 3c^2 d^2 \left(\frac{d\varepsilon_n}{d\lambda}\right) \cdot \left(\frac{d^2 \varepsilon_n}{d\lambda^2}\right) - c^3 d^3 \left(\frac{d\varepsilon_n}{d\lambda}\right)^3. \tag{50}$$

Higher derivatives can be formed accordingly.

Compared with the derivatives of absorbance A with respect to wavelength (Eq. (47)), the derivatives of light intensity I with respect to wavelength λ are more complicated, but we can see important correlations more clearly. Whilst the first derivative is directly proportional to the concentration in both cases, this no longer applies directly to the 2nd and 3rd derivatives as can be seen immediately from Eqs. (49) and (50).

In the case of the 2nd derivative (Eq. (49)), the first derivative of ε must equal zero, $d\varepsilon_n/d\lambda = 0$ in order to maintain a linear proportionality with the concentration. This will be the case at a band maximum where the 2nd derivative also assumes an extreme value and $d^2I/d\lambda^2$ becomes particularly large. This is associated with an especially high sensitivity of measurement.

The 1st derivative always acquires maximum values at the inflexion points of the zero-order spectrum since $dA/d\lambda$ or $d\varepsilon_n/d\lambda$ are at their maxima there. Thus, the most favorable analytical positions can be chosen by reference to all the Eqs. (48) to (50).

The simplest method of obtaining first and second order derivative spectra consists of electronic differentiation which is most commonly used today and it is also economical in its implementation.

If λ_1 and λ_2 are selected so as to be very close in a dual-wavelength spectrometer (see Sect. 5.1) the 1st derivative is obtained directly from $\Delta A/\Delta\lambda$ [1]. However, the 2nd derivative must be obtained by electronic means.

Another possibility of obtaining derivative spectra is provided by the *self-modulation method* developed by Bonfiglioli, Brovetto et al. [13] and the *wavelength modulation method* [14]. In most cases, derivative spectra are produced by an electronic differentiation as mentioned above. In most UV-VIS spectrophotometers fitted with a microcomputer, a program to form the 1st and 2nd derivatives as well as higher derivatives is part of the standard equipment. The derivatives are obtained digitally by calculation from the individual experimental data points.

Although we can usually work adequately with first and second order derivative spectra, it may be desirable to generate higher order derivative spectra. Talsky et al. have investigated this in detail [15–19]. They have reviewed electronic methods for higher order differentiation. For derivatives $n > 3$ only analog and digital computers are suitable [15] but analog on-line processing is preferable on account of the computation time. Talsky has designed analog differentiators for practical applications which permit production of derivative spectra up to the sixth or ninth order. Although information is lost at every differentiation, higher order derivative spectra can nevertheless provide useful information in many cases.

The example in Fig. 27 illustrates the typical properties of a first and second order derivative spectrum. The 1st derivative represents the slope at each point of the absorption band. $dA/d\lambda$ in Fig. 27 has its maximum value at the turning point of the absorption band on the shorter wavelength side and its minimum value at the turning point on the long wavelength descending flank of the absorption band. $dA/d\lambda$ itself equals zero at the absorption maximum. Since $dA/d\lambda$ can be greater or smaller than zero we set half of the full-scale deflection of the absorbance scale to $A = 0.5$ in Fig. 27. At the absorption maximum λ_{max} $dA/d\lambda = 0$. For broad maxima this means that we can determine λ_{max} very accurately from the point of intersection of curve $dA/d\lambda = f(\lambda)$ with the zero axis. For the 2nd derivative in Fig. 27, the values of $d^2A/d\lambda^2$ are zero at the inflexion points of the absorption spectra according to the rules of differentiation. At the position of λ_{max}, a minimum value is obtained and this again permits the location of the absorption maximum very accurately.

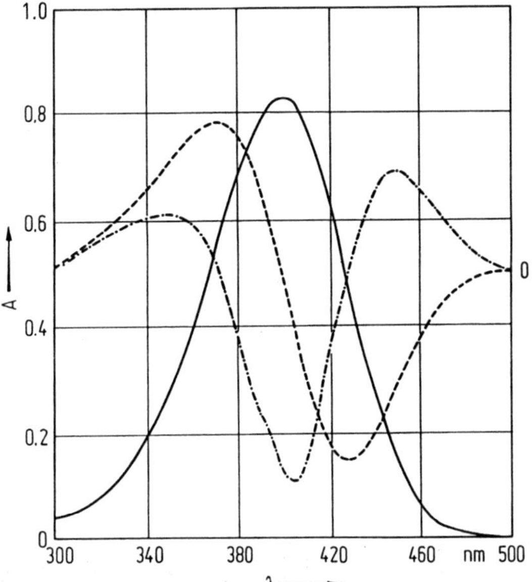

Fig. 27. Derivative spectra: first order (– – –), second order (–.–.–) and zero order absorption spectrum (———)

Thus, we have established that the zero crossings of first order absorption spectra and the minima of second order absorption spectra correspond to the appropriate absorption maxima.

Since $dA/d\lambda$ or $d^2A/d\lambda^2$ are extremely sensitive to any change of the slope in an absorption spectrum this method is very well suited for analyzing shoulders and overlapping absorption bands [12, 16].

The method is also very important to molecular spectroscopists, particularly for the analysis of the structure of absorption bands. In order to illustrate this, Fig. 28 shows the zero and second order spectra of *pyridine in n-heptane* [20]. Whilst the structure of the absorption band of pyridine is only weakly outlined in the zero order spectrum (standard absorption spectrum), the second order spectrum permits an unambiguous analysis of the vibrational structure of these absorption bands.

Derivative spectroscopy has an important application in analysis; not only for the detection of trace elements but also for quantitative determination. Schmitt [21] has reviewed numerous applications. O'Haver and Green [2] have detailed the errors involved in applying this method to the quantitative analysis of mixtures.

Figure 29 shows an example of the concentration dependence of the zero, first and second order spectra [23]. The zero, first and second order spectra of *isoquinoline in n-heptane solution* are shown on the left hand side of the figure.

The relative intensities of these spectra at the wavelengths denoted by λ_1 to λ_4 at different isoquinoline concentrations are plotted on the right hand side.

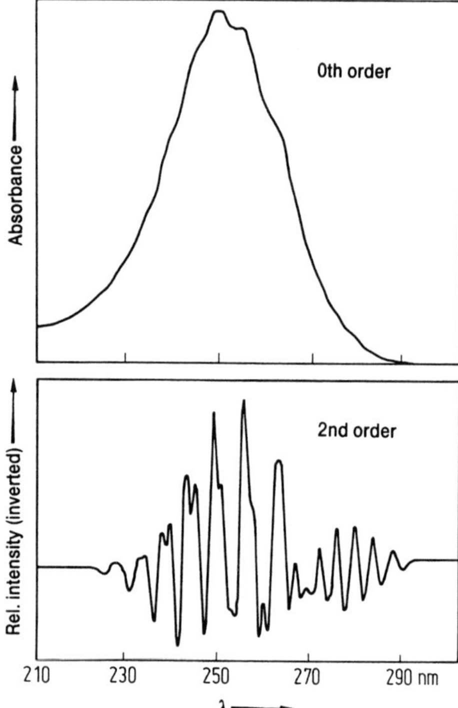

Fig. 28. Absorption spectrum and second order derivative spectrum (inverted) of pyridine in *n*-heptane. Inverted means that the negative values of the absorbance of the 2nd derivative have been plotted in the direction of the positive absorbance values of the absorption spectrum. Thus, the maxima of the inverted 2nd derivative correspond to the absorption maxima

The determination of concentration by means of a first order spectrum has many advantages when *investigating mixtures*. We take an example where one of the components in a mixture has characteristic vibrational fine structure $A_a(\lambda)$ in a specific wavenumber region whilst the other components only provide the (practically) constant absorbance contribution of $A_b(\lambda) = \text{const}$ in this wavenumber region.

Here, the total absorbance $A_{tot}(\lambda)$ is given by:

$$A_{tot}(\lambda) = A_a(\lambda) + A_b(\lambda) + \ldots = A_a(\lambda) + \text{const} \ , \tag{51}$$

$$dA_{tot}/d\lambda = dA_a(\lambda)/d\lambda + dA_b(\lambda)/d\lambda = A'_a(\lambda) + 0 \ . \tag{51a}$$

In the first order spectrum, the constant contribution $A_b(\lambda)$ disappears and the concentration of component a can be determined directly from a first order spectrum by means of a calibration line. It can be seen immediately in Eq. (51) that, in this case, the loss of information associated with a differentiation has a positive effect since the interfering background is eliminated [17].

The determination of concentration by means of first and second order spectra can be successfully applied in *turbid solutions* (constant background: see above) and in particular in biological systems [12, 14, 18, 24, 25].

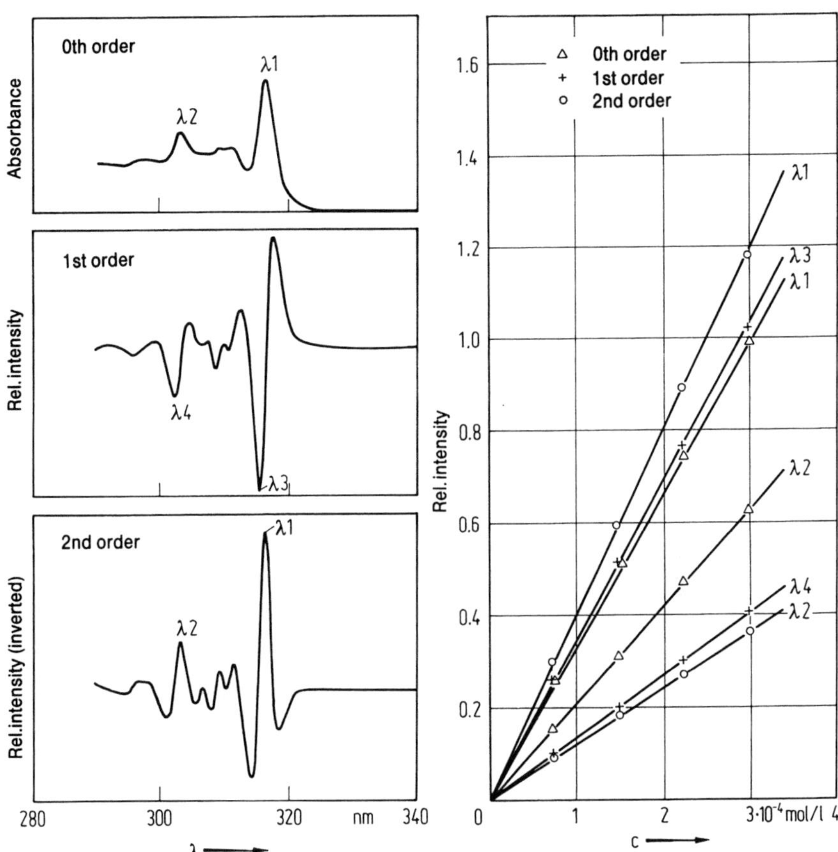

Fig. 29. Illustration of the concentration dependence of the zero, first and second order spectra of isoquinoline in *n*-heptane

The water/phenol system may be used as an example of the analysis of a turbid solution by means of derivative spectroscopy. In turbid solutions such as in *industrial waste water*, the background is strongly pronounced due to scattering. Therefore, the quantitative determination of phenol is subject to a large error. However, since this background shows a continuous increase toward shorter wavelengths we can eliminate it almost completely by forming $dA_s/d\lambda$ or $d^2A_s/d\lambda^2$ (see Eqs. (51, 51a)) as shown by Shibata et al. [26, 27].

As illustrated here, this technique can be used for other systems, particularly for multicomponent analyses and association equilibria [20, 23]. Binary mixtures of substances, whose absorption spectra differ very little or hardly at all, can be determined qualitatively and quantitatively by means of their derivative spectra. This has been demonstrated with mixtures of 2,4-dichlorophenol and 2,4,6-trichlorophenol [27].

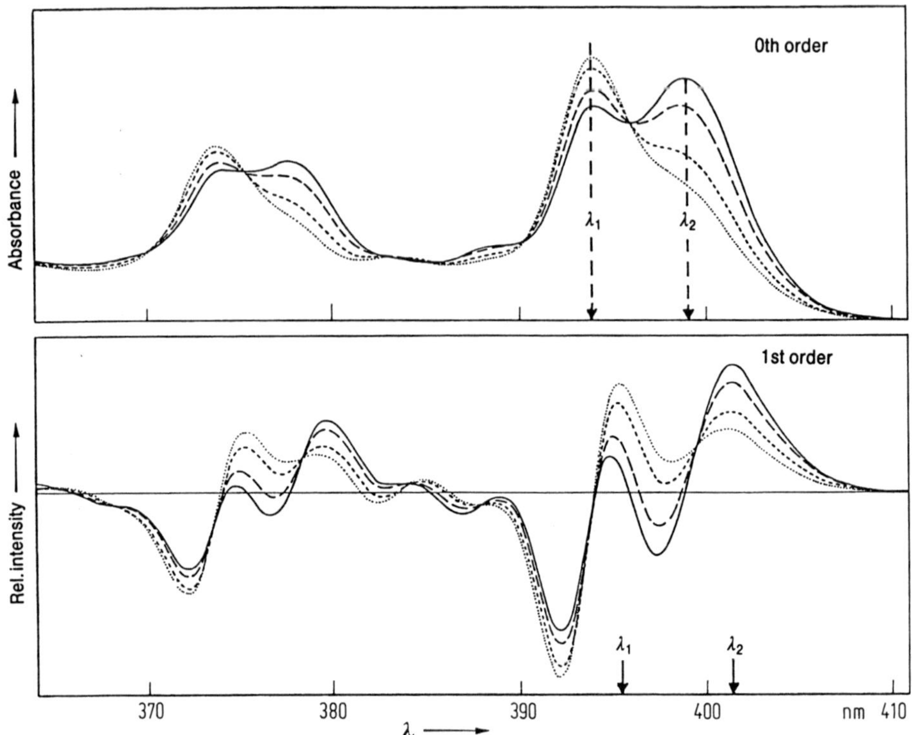

Fig. 30. 1,2,7,8-dibenzacridine/4-bromphenol in toluene at different temperatures; absorption spectra and first order spectra; ——— 275 K, – – – 282 K, ------ 294 K, ... 303 K

Another example of a system where the absorption spectra of two components show little difference are *hydrogen bonding association equilibria* in the UV-VIS spectral region. The absorption spectrum of an H-bonded complex of 1,2,7,8-dibenzacridine with *p*-bromphenol in *n*-heptane is shifted bathochromically by 250 cm^{-1} vis-a-vis the spectrum of the pure base in *n*-heptane. The vibrational structure is maintained in the complex [23]. If the first derivative spectra of such a system are examined, as shown in Fig. 30, we see that wavelengths λ_1 and λ_2 can very easily be found at which complex (λ_1) does not underlie the free base and the free base (λ_2) does not underlie the complex. This is in contrast to the case for the 0th order spectra [23]. Consequently, the concentration of uncomplexed base in an equilibrium mixture can be determined without difficulty at analytical position λ_1 by means of a suitable calibration plot.

Numerous examples show that there is comprehensive scope for applications in environmental analysis, food, clinical and physiological chemistry as well as in biochemistry [15, 17, 18, 21, 28, 29, 30, 31].

5.3 Reflectance Spectroscopy

The Bouguer-Lambert-Beer law presupposes samples in which the light intensity is not lost by scattering and reflection processes. In molecularly dispersed systems, the scattering losses due to particles are so small that they lie well below the photometric accuracy. Reflection losses occurring at every phase boundary are eliminated in practice by measuring versus a reference cuvette. However, the situation is quite different in the case of samples which scatter strongly or are opaque to light since in this case the incident light is reflected diffusely. The reflecting power is also a function of the absorbing power of a substance. The recognition of the color of a substance is based on these facts. The complementary color is absorbed by the substance while the eye perceives the radiation which is not absorbed.

Since diffuse reflection is caused by single and multiple scattering on the surface of and inside a solid substance, the remitting power of substance can be represented to a first approximation as a function of the absorption coefficient (β in cm^{-1}) and scattering coefficient (s in cm^{-1}). This *two-constant theory* led Kubelka and Munk [32] to deal with this problem theoretically. We have to thank Kortüm for a discussion of the basic principles and applications [33].

In reflectance spectroscopy, the Kubelka-Munk function $F(R_\infty)$ replaces the Bouguer-Lambert-Beer law. This function establishes a correlation between the *diffuse reflecting power* R_∞, the absorption coefficient $K = 2\beta\,(cm^{-1})$ and the scattering coefficient $S = 2\,s\,(cm^{-1})$ of a sample:

$$F(R_\infty) = \frac{(1-R_\infty)^2}{2R_\infty} = \frac{K}{S} \,. \tag{52}$$

Within the scope of the theory, R_∞ implies that the sample thickness approaches infinity ($d = \infty$) while the background reflectance is simultaneously zero ($R_g = 0$). The factor 2 in the absorption and scattering coefficients K and S, as defined by the Kubelka-Munk theory, can be attributed to the fact that the radiation flux of the incident and scattered light in both directions in the sample must be considered [34].

The scattering coefficient S and absorption coefficient K for finite pathlength d are given by:

$$S = \frac{2.303}{d} \cdot \frac{R_\infty}{2-R_\infty^2} \cdot \log \frac{R_\infty(1-R_0 \cdot R_\infty)}{R_\infty - R_0} \,, \tag{53}$$

$$K = \frac{2.303}{2d} \cdot \frac{1-R_\infty}{1+R_\infty} \cdot \log \frac{R_\infty(1-R_0 \cdot R_\infty)}{R_\infty - R_0} \,. \tag{54}$$

In these equations, R_0 is the diffuse reflecting power of a sample in front of an ideal black non-reflecting background for which, as for $d = \infty$, the reflectance R_g equals zero. Equations (53) and (54) again yield the Kubelka-Munk function, Eq. (52) directly.

Equation (55) provides the correlation between the diffuse reflecting power for the different cases:

$$R_0 = \frac{R_\infty (R_g - R)}{R_g - R_\infty (1 - R_g R_\infty + R_g R)} \tag{55}$$

where

R_g is the reflectance of the background for $d = 0$,
R is the reflectance of the sample for $d > 0$,
R_∞ is the reflectance of the sample for $d = \infty$,
R_0 is the reflectance of an ideal black non-reflecting background

when $d = \infty$, because $R_g = 0$, and $R = R_\infty$, $R_0 = R_\infty$.

In practice, the relationship $R_0/R(R_g)$ is frequently used to characterize a diffusely reflecting layer; and this relationship can have different values depending on the reflecting power R_g of the background.

At $R_g = 1$, an *ideal* white background, we describe $R_0/R(R_g = 1)$ as an ideal *contrast relationship* which cannot be measured in practice because $R_g = 1$ cannot be realized. For that reason, the relationship $R_0/R(R_g = 0.98)$ is generally used. In this case, freshly deposited MgO or TiO_2 is utilized as a white background.

The quantity R_∞ is important in the application of reflectance spectroscopy but it cannot be measured accurately with conventional equipment. Therefore, the diffuse reflectance R_∞ is always related to a white standard as reference, i.e. it is obtained as a relative value, R'_∞:

$$R'_\infty = \frac{R_{sample}}{R_{standard}}. \tag{56}$$

If the absolute reflecting power ϱ of a white standard R_{St} could be made equal to 1, then the absolute and relative reflecting power of a sample would be the same. However, there is no known white standard which shows this property over the total spectral region (UV-VIS-NIR) of interest.

Therefore, the absolute reflecting power, R_∞, of the standard must be known in order to determine R_{sample}.

In practice, *deposited MgO* has proved to be the best standard on account of its simple production under defined conditions. For that reason, many measurements have been carried out in order to determine the absolute reflecting power ϱ of MgO as a function of the wavelength. In the visible spectral region, the ϱ-values are 0.983 ($\lambda = 420$ nm) and 0.986 ($\lambda = 680$ nm) with a maximum value of 0.988 ($\lambda = 620$ nm) [35]. Kortüm et al. have determined the absolute reflecting power for frequently used white standards which are of interest for specialist physico-chemical applications [36]. In addition to MgO the following substances have been measured in the UV-VIS and NIR region:

Li_2CO_3; NaF; NaCl; $MgSO_4$; $BaSO_4$, aerosil; Al_2O_3, SiO_2 and glucose.

The results show that the absolute reflecting power of these materials decreases strongly towards the UV region [33]; and the same applies to the NIR region. Aerosil is an exception in that it has values between 0.90 and 0.99 above 30000 cm^{-1}.

One of the methods for determining the absolute reflecting power ϱ is based on the application of Eq. (55). If relative values, which are related to the same white standard, are introduced into this equation i.e.

$$R' = \frac{R}{\varrho}, \quad R'_g = \frac{R_g}{\varrho}, \quad R'_0 = \frac{R_0}{\varrho} \quad \text{and} \quad R'_\infty = \frac{R_\infty}{\varrho},$$

an expression is obtained from Eq. (55) which can be solved for ϱ. Therefore, the absolute reflecting power of a standard can be accessed via the relative measurements [37]. The most common method is based on the application of the Taylor-sphere theory described in detail by Kortüm [33] who also discusses other methods in his book.

Equation (56), which defines the relative diffuse reflecting power, has a formal similarity with the definition of transmission in standard absorption measurements. There the weakened intensity I after traversing the sample can be related to the unimpaired intensity I_0 after traversing a reference cuvette. However, in reflectance spectroscopy, the diffuse reflecting power is always related to that of a white standard. Basically, this determines the design of the reflectance accessory fitted to most spectrophotometers and the technique of measurement itself.

A *photometric sphere* coated on the inside with either MgO or $BaSO_4$ is an essential component of a reflectance accessory. This sphere has the task of integrating the diffusely reflected light from sample and standard. Therefore, this device is called on integrating sphere accessory for diffuse reflectance.

Figure 31 shows an outline of an integrating sphere accessory for measurements in single-beam mode. The light emerging from the exit slit of a monochromator is focussed onto the sample via the lens and mirror. The light diffusely reflected from the sample is collected by the sphere and passes to the multiplier. The measured signal is proportional to the diffuse reflecting power of the sample.

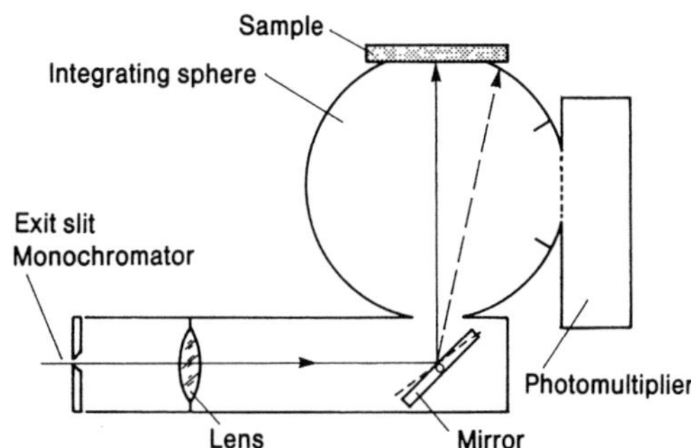

Fig. 31. Integrating sphere accessory for a single-beam spectrophotometer e.g. Zeiss PMQ III

If we rotate the mirror slightly (dotted line) when making the second measurement, the light falls onto a position on the internal wall of the integrating sphere which, at the same time, serves as white standard. Therefore, the diffusely reflected light supplies the reference signal. If the ratio of two signals is formed, the relative diffuse reflecting power R'_∞ of the sample at a given wavelength, λ, is obtained directly. We call this measurement geometry ${}_0R_d$:

Left subscript: *zero* degree (normal) irradiation,
Right subscript: *diffusely* measured reflectance.

However, it is frequently more appropriate to work with a moveable sample holder which permits the positioning of a sample and standard consecutively in the same position (here the mirror does not have to be rotated).

The experimental arrangement shown in Fig. 31 can be reversed, i.e. we irradiate diffusely with white light and focus the light reflected from the sample onto the entry slit of the monochromator. The corresponding geometry is denoted as ${}_dR_0$. This can be easily achieved by substituting a continuous source of light for the photomultiplier shown in Fig. 31. We must ensure by means of appropriate appertures that no direct light from the source falls onto the sample and standard.

In modular single-beam photometers such as the Zeiss PMQ III, the setup can be changed without difficulty. Spectrophotometers which can be converted or extended into spectrofluorimeters should make geometry ${}_dR_0$ possible since the measurement of a reflectance spectrum corresponds to that of a fluorescence spectrum.

Figure 32 shows schematically the light path in a double-beam spectrophotometer with an integrating sphere accessory (simplified reproduction of the light path in the accessory for the Perkin-Elmer 55X series spec-

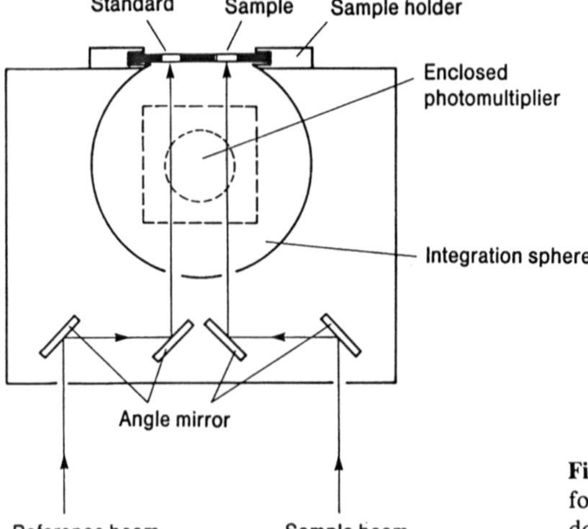

Fig. 32. Integrating sphere accessory for the Perkin-Elmer model 555 double-beam spectrophotometer

trophotometers). The auxiliary equipment is fitted with a fixed sample and cuvette holder. The fixed sample holder accepts samples of minimum dimensions 12 mm×22 mm up to a maximum of 40 mm×40 mm by 6 mm thick. This accessory also provides the measurement geometry ${}_0R_d$.

The photomultiplier (dotted line) is fitted on the sphere surface vertically above the paths of the light beams. With this accessory and extensions, measurements on turbid solutions and transparent solids using transmission techniques can be made.

The Kubelka-Munk function, Eq. (52) applies only to diffuse reflectance. Considerable deviations can occur as soon as contributions from regular (specular) reflection intrude. By using the *dilution method,* this contribution can be eliminated. We dilute the powdered substance under investigation with an inert, nonabsorbing solid standard (MgO, NaCl, $BaSO_4$, SiO_2, TiO_2, etc.) to such an extent that the specular contribution of the remittance in the relative measurement against the same pure standard is reduced to within the accuracy of measurement of the method [38]. The intimate mixing required is achieved by pulverizing or by grinding the sample in a ball mill. Either a simple homogeneous mixture of crystallites is obtained or the sample is absorbed as a molecular dispersion on the surface of the standard. This is usually the case when organic solid substances are ground with inorganic standards. In such instances, the reflectance spectrum of the absorbed substance is measured. These cases are of particular practical interest and it has been shown that absorption coefficient K in Eq. (52) is proportional to the concentration of the adsorbate [39]. This means that the scattering coefficient is constant for a series of dilutions with the same standard.

Thus, the Kubelka-Munk function depends only on the absorption coefficient which can then be formulated as a product of the molar decadic extinction coefficient ε and the concentration c. Equation (52) becomes:

$$F(R_\infty) \cong \frac{\varepsilon \cdot c}{S}$$

or

$$\log F(R_\infty)_\lambda = \log \varepsilon + C \ ,$$
$$C = \log \frac{c}{s} = \log c + \text{const} \ . \tag{57}$$

This correlation has been confirmed in many systems and many applications of reflectance spectroscopy are based upon it. Equation (57) corresponds to the Lambert-Beer law for transmitted light measurents. Both laws apply in the limited area of high dilution. Since the Kubelka-Munk function depends on the wavelength, the representation of $\log F(R_\infty)$ as a function of the wavelength reproduces the absorption spectrum in the form of a *typical color curve.* A parallel shift of the ordinate brings this curve

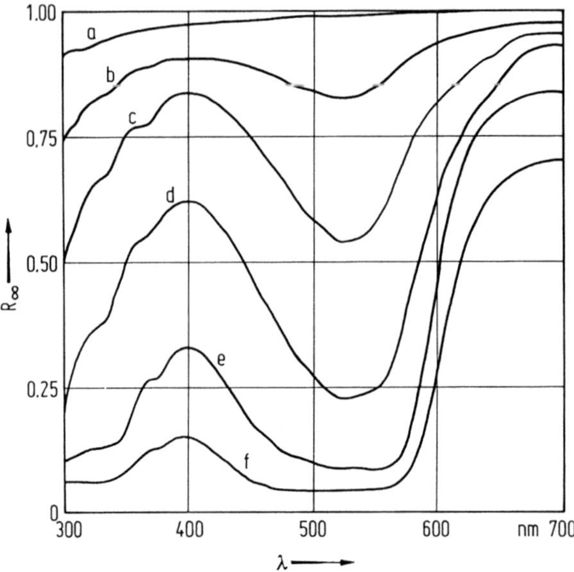

Fig. 33. Reflectance spectra of a series of dilutions of an organic pigment dye (Hostaperm red E3B) with MgO. *a* MgO, *b* pigment = 1 : 1000, *c* 1 : 100, *d* 1 : 10, *e* 1 : 1, *f* full-tone pigment

into coincidence with the true absorption spectrum measured in transmitted light. This simple correlation between a typical color curve in the reflectance spectrum and the true spectrum applies only if the scattering coefficient is independent of the wavelength and if the standard itself has no absorption.

Figure 33 shows the reflectance spectra of an *organic pigment dye* as a typical example of the dilution method. This organic pigment has a high absorption coefficient. Consequently, no spectral information can be gained from the reflectance spectrum of the pure (undiluted) material.

However, on dilution of the pigment with MgO, the characteristic absorption curve can be clearly seen (cf. Fig. 33). This example of a solid substance with a high absorption coefficient also shows that it is frequently impossible to determine the light absorption of pure dyes by reflectance spectroscopy.

As shown in Eq. (57), if $\log F(R_\infty)$ is plotted against the logarithm of a concentration (dilution) or $F(R_\infty)$ against the concentration, a linear correlation is obtained (see examples in [33]).

These applications of reflectance spectroscopy are of particular interest when we can obtain a solution only with difficulty or not at all. To this category belong the spectra of insoluble substances or substances which react in solution, spectra of absorbed substances, kinetics measurements, spectra of crystalline powders, dynamic reflectance spectroscopy, analytical photometric measurements as well as the measurement and matching of

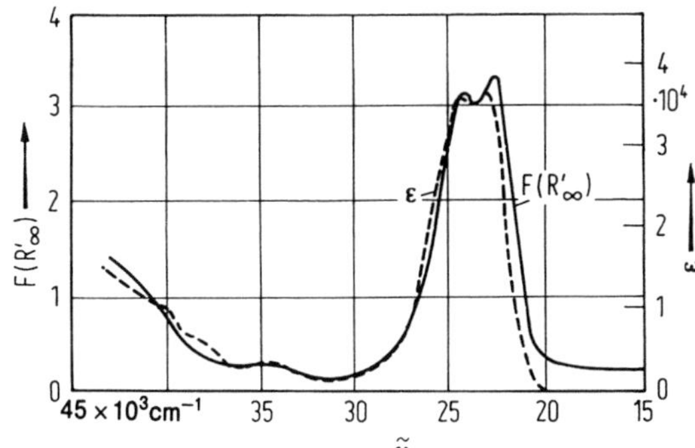

Fig. 34. Reflectance spectrum of triphenylchloromethane absorbed on SiO_2/Al_2O_3 (———) and the absorption spectrum in H_2SO_4 (– – –)

color [33]. The application to analysis, and in particular to environmental problems, has been described by Frei and McNeil [40].

An interesting example resulting from investigations by Kortüm and Friz is shown in Fig. 34 [41]. The figure shows the reflectance spectrum of triphenyl chloromethane adsorbed on a SiO_2-Al_2O_3 cracking catalyst. Comparison with the absorption spectrum of triphenyl chloromethane in concentrated H_2SO_4 shows that adsorption of this compound on an acidic oxide surface gives rise to the triphenyl methyl cation. Benzyl chloride and diphenylmethyl chloride behave analogously. On basic oxide surfaces such as MgO, the adsorption also results in a triphenyl methyl cation which then reacts further on the surface. The example shows the possibilities for the use of reflectance spectroscopy in the investigation of heterogeneous catalysis extremely well.

The quantitative photometry of thin-layer chromatograms provides an analytical application. Equipment has been developed in recent years which permits routine analysis. Hezel [42, 43] has written a review. Messrs Carl Zeiss have published an extensive list of references for the period 1966–1977 [44].

5.4 Photoacoustic Spectroscopy

5.4.1 Principles of PAS

As early as 1880/81, *Alexander Graham Bell* discovered the photoacoustic effect (PAE) and the photoacoustic spectroscopy (PAS) of solid substances derived from it [45]. At approximately the same time, *Tyndall* and *Röntgen* observed the effect in gases [46, 47]. According to Bell, an absorption process is the primary step for generating the PA-effect. An acoustic signal, i.e. a sound wave is observed. We are dealing with the transformation of absorbed light energy into mechanical energy.

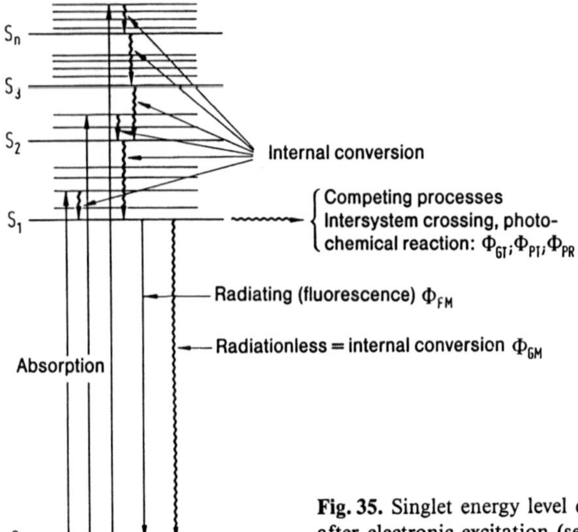

Fig. 35. Singlet energy level diagram of primary processes after electronic excitation (see explanation in text)

The energy level diagram in Fig. 35 provides an explanation for this. The interaction of a molecule with electromagnetic radiation is primarily an absorption which leads in $10^{-15}-10^{-14}$ s to excitation to the states S_1, $S_2 \ldots S_n$ including their accompanying vibrational levels. From the higher singlet states $S_2, \ldots S_n$ a radiationless deactivation to S_1 results within $10^{-13}-10^{-12}$ s. In S, the excitation energy previously stored in $S_2, \ldots S_n$ is converted into heat by vibrational relaxion processes (*internal conversion*). After $10^{-9}-10^{-8}$ s, the S_1 state is also deactivated which can happen either:

1. by fluorescence or
2. nonradiatively by conversion of the excitation energy $S_0 \rightarrow S_1$ into heat (internal conversion)

If light of a specific modulation frequency impinges on such a closed system, heat generation will accompany the modulation frequency in a periodic manner, provided that the frequency is low compared with the speed of the deactivation processes,

1. internal conversion
2. competing processes
   intersystem crossing
   photochemical reaction: Φ_{GT}; Φ_{PT}; Φ_{PR}
3. radiative transition (fluorescence) Φ_{FM}
4. radiationless transition = internal conversion Φ_{GM}
5. absorption

In the case of a gas in a closed system (V = const) the pressure of a gas changes periodically and the generated pressure wave corresponds to a

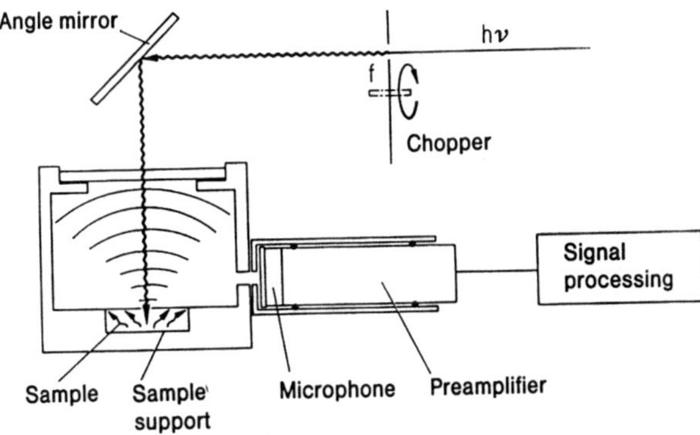

Fig. 36. Outline of a photoacoustic cell for solid substances

sound wave. If a microphone is fitted to the cell wall this acoustic signal can be recorded directly.

In condensed samples such as liquids, solutions and solid substances matters are more complicated, cf. Fig. 36. A hole for connecting a microphone is drilled into the side wall of a cell for solids. It is essential that a *gas layer* lies above the sample. Light modulated at frequency f is directed through a quartz window via an angle mirror and is absorbed by the sample. Subsequently, the deactivation processes described take place and during their course heat is released very rapidly. After local heating inside the sample, heat diffuses to the sample surface. At the sample/gas phase boundary the *heat wave* enters the gas and becomes a pressure wave, i.e. a sound wave which can be recorded with a microphone. Such an arrangement is called a *gas-coupled photoacoustic cell.*

Since samples are deposited on a support we must take account of the thermal properties of support and sample as well as of the coupled gas. Theoretical considerations [48, 49] show that the heat wave excites only a thin gas layer at the sample/gas interface whose thickness changes periodically with modulation frequency $\omega = 2\pi f$. It can be regarded as a moving piston which transmits the pressure wave adiabatically to the remaining gas volume.

In PAS, the absorption of light is related to the absorption coefficient β:

$$A_{\bar{\nu}} = \ln \frac{I_0}{I} \bar{\nu} = 2{,}303 \cdot \varepsilon \cdot c \cdot d \ , \quad \beta = \frac{A_{\bar{\nu}}}{d} = 2{,}303 \cdot \varepsilon \cdot c \ . \tag{58}$$

The absorption coefficient has the dimension "cm^{-1}". If the measured PAS signal (S^{PA}) is directly proportional to the absorption coefficient β we can expect to obtain a *photoacoustic spectrum*, corresponding to the well-known absorption spectrum, by plotting S^{PA} as a function of the wavelength (wavenumber).

However, in reality considerable deviations occur as can be seen from Fig. 35. Deactivation of the higher excited states $S_2, S_3 \ldots S_n$ to state S_1

generally occurs by conversion to heat by *internal conversion.* This state (S_1) has a long life compared with the S_2, $S_3 \ldots S_n$ states.

Subsequently, several deactivation mechanisms can occur, but account must be taken of their quantum yields:

1. radiative deactivation = fluorescence Φ_{FM}
2. intersystem crossing to the triplet state with Φ_{ISC}
   a) radiative deactivation = phosphorescence Φ_{PT}
   b) nonradiative deactivation Φ_{GT}
3. photochemical reaction and Φ_{PR}
4. nonradiative deactivation $S_1 \rightarrow S_0$ Φ_{GM}

Processes 2b) and 4 are desirable in PAS. All other deactivation mechanisms compete with PAS and do not contribute to a PA-signal.

At ambient temperature it can be generally assumed that the processes under 2 do not occur, i.e. Φ_{ISC}, Φ_{PT} and $\Phi_{GT} = 0$. If, in addition, systems are investigated in which no photochemical reactions occur under the given circumstances, then

$$\Phi_{GM} = 1 - \Phi_{FM} \ . \tag{59}$$

Frequently, we cannot ignore the fluorescence, and so, it competes most strongly with the PA-effect. However, Eq. (59) provides us with the opportunity of determining fluorescence-quantum yields absolutely.

Rosencwaig and Gersho have developed a theory of the generation of a PA-signal in condensed phases. They used the one-dimensional model outlined in Fig. 37 as a basis [48].

An optically and thermally homogeneous sample with pathlength l_s is placed on a support with a thickness of l_u in a cylindrical PA-cell. The sample is illuminated with sinusoidally modulated light.

The energy absorbed by the sample is completely deactivated nonradiatively. We consider only incident light and resultant heat flow perpendicular to the sample surface. The Bouguer-Lambert-Beer law applies to the absorption of electromagnetic radiation in a sample. Taking account of the resultant heat distribution, the thermal diffusion equations for the sample, sample support and gas phase can be obtained.

A theoretical calculation of the PA-signal is possible by complete solution of the Rosencwaig-Gersho (RG) equations [48, 49] if we know all the

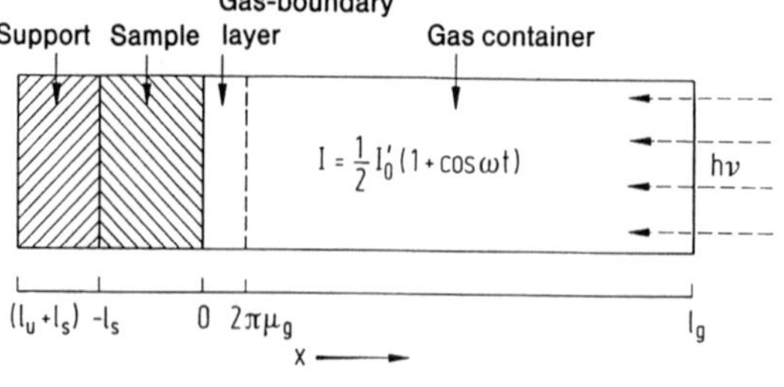

Fig. 37. Schematics of the model for the Rosencwaig-Gersho theory and parameter definitions

optical and thermal data and the radiation flux. Nevertheless, an analysis of limiting cases has proved appropriate in practice for the investigation of solid substances and liquids when using the RG theory.

In the first instance, we define the optical penetration depth l_β and the thermal diffusion length μ_s:

$$l_\beta = \frac{1}{\beta} \text{ in cm} \quad \text{and} \quad \mu_s = \sqrt{\frac{2\alpha_s}{\omega}} \text{ in cm} \tag{60a, b}$$

l_β is the reciprocal value of the absorption coefficient. Since $I = I_0 e^{-\beta l_\beta}$, the intensity I_0 has dropped to the value of $I = 0.37\ I_0$ at depth $l = l_\beta$.

The thermal diffusion length μ_s is a function of the thermal diffusivity and of the modulation frequency. The latter is a very important factor.

Taking account of the pathlength of sample l_s we can differentiate between:

1. optically transparent $l_\beta > l_s$
2. optically opaque $l_\beta < l_s$

In both cases we can also differentiate between:

thermally thin samples $\mu_s > l_s$
thermally thick samples $\mu_s < l_s$

However, of all the possible combinations only two are of practical importance, i.e.:

1. optically transparent and thermally thick $l_\beta > l_s$
2. optically opaque and thermally thick $l_\beta < l_s$

As far as option 2 is concerned we must differentiate between:

2a) thermally thick with $\mu_s > l_\beta$
2b) thermally thick with $\mu_s < l_\beta$

The same solution for the PA-signal is found for cases 1 and 2b

$$Q^{PA} = -i \frac{\mu_g I_0}{2} \frac{\mu_s}{\varkappa_s} (\beta \mu_s) \cdot B \tag{61}$$

with

$$B = \frac{\gamma P_0 \eta_s}{2\sqrt{2} T_0 V}$$

P_0: external pressure, γ: $(C_p/C_v)_{gas}$, T_0: external temperature, V_0 volume of the PA-cell, η_s: conversion coefficient.

Case 2a) yields the expression:

$$Q^{PA} = (1 - i) \cdot \frac{\mu_g I_0}{2} \frac{\mu_s}{\varkappa_s} \cdot B \ . \tag{62}$$

A comparison of Eqs. (62) and (61) shows that the photoacoustic signal is independent of the optical properties of the sample in this case, i.e. there is *signal saturation*. These cases of thermally thick samples can be summarized for any optical density [50, 51]:

$$Q = \frac{\beta(r-1) I_0 \mu_s}{\varkappa_s (g+1)(\beta^2 - \sigma^2)} B . \tag{63}$$

The quantities g, r and σ are defined as follows [48, 49]:

$$g = \frac{\varkappa_g}{\varkappa_2} \sqrt{\frac{\alpha_s}{\alpha_g}} , \quad \sigma = (1+i) \sqrt{\frac{\omega}{2\alpha_2}} , \quad r = (1-i) \frac{\beta}{2} \sqrt{\frac{2\alpha_s}{\omega}} .$$

Equations (61) to (63) are complex functions which contain real and imaginary parts. From the complex expressions the amplitude of the PA-signal q is obtained as:

$$q = \sqrt{Q_{real}^2 + Q_{im}^2} .$$

For phase angle Φ the result is:

$$\Phi = \arctan (Q_{im}/Q_{real}) .$$

The PA-signal, Q^{PA}, is initially represented as pressure units (Pa) in all equations. The real PA-signal amplitude $q = |\Delta p|$ must be converted into volts for practical purposes, i.e. the sensitivity of the microphone, S^M, must be known in mV/Pa. Then, the PA-signal, S^{PA}, is obtained in mV as:

$$S^{PA} = \Delta p \, Q_{aku}(\omega) S^M(\omega) .$$

$Q_{aku}(\omega)$ takes account of the fact that the PA-measuring cell is a Helmholtz resonator whose resonance frequency depends upon the chopper frequency $f = \omega/2\pi$ and influences the sensitivity of the gas-coupled detection. For measurements made at constant frequency f, $Q_{aku}(\omega)$ is always constant. If the amplitude of the PA-signal S^{PA} is of interest then the following correlation between the intensity-corrected PA-signal and absorption coefficient can be derived from Eq. (63) and the definitions of g, r and σ:

$$S^{PA} = \frac{\beta \left[\frac{\beta \mu_s}{F} - 1 \right]}{\beta^2 - \frac{2}{\mu_s^2}} F . \tag{64}$$

Factor "F" contains, in addition to the parameters contained in g, r and σ, the microphone sensitivity (see above), the constant parameter B, see Eq. (61) and radiation intensity I_0.

F remains constant during the measurement of a PA-spectrum. If the thermal diffusion length, μ_s, and absorption coefficient, β, at a given wavelength, which is the reference point, are known, the factor F in Eq. (64) can be determined from the appropriate PA-signal. Equation (64) can then be used to assign absorption coefficients (cm^{-1}) to the corrected PA-signals (arbitrary units) [51, 53]. As a result of the saturation effect, this leads to a *nonlinear scaling* of the ordinate in the region of larger absorption coefficients.

Only on the assumption that μ_s is less than $1/\beta = l_\beta$ is a linear correlation between the PA-signal and the absorption coefficient obtained, see Eq. (61). Equation (64) can then be written in the simplified form:

$$S^{PA} = \frac{\beta \mu_s^2}{2} F \; . \tag{65}$$

If the constants, or the parameters kept constant, which occur in F, are included, then this corresponds to the limiting case of Eq. (61).

McDonald and Wetsel extended the concept of a "thermal piston" used by the RG theory [53]. In addition to the heat conductance characterized by the RG theory, acoustic waves generated by thermoelastic effects in the sample contribute to the signal (Composite Piston (CP) model).

Burggraf and Leyden [54] have proposed an extension of the RG theory for strongly light-scattering thermally thick samples. Starting from continuum mechanics, Pelzl and Bein have derived a general equation to describe the generation of acoustic signals for solid substances which was used to investigate the theoretical ideas outlined here [55].

In case 2a) we illustrated by means of Eq. (62) that there is signal saturation, i.e. that the PA-signal is independent of the absorption coefficient β of the substance under investigation. Thus, the PA-signal is then only a function of radiation intensity I_0. Since I_0 of the light source shows a spectral intensity distribution, the PA-signal as a function of the wavelength would then correspond to the spectrum of the light source.

A substance which absorbs all of the penetrating light completely over a broad spectral region, i.e. has a very large β-value in this region, can be regarded as an ideal black body. In PAS, we can approximate such a body by a *carbon standard* which can easily be produced by blackening a glass or quartz plate. The PA-spectrum of this standard would then correspond to the combination of lamp plus monochromator.

The saturation effect of a *black standard* can be used to generate a reference signal for PA-measurements. Therefore, we can relate the signal of any sample at the same wavelength to radiation intensity I_0. This property is extremely important when designing a double-beam PA-spectrophotometer.

In all other substances which are investigated with PAS, the saturation effect is an undesirable phenomenon which is found frequently in strongly absorbing systems.

The requirements for limiting case 2a) in Eq. (62) show that the thermal diffusion length μ_s is greater than the optical penetration depth l_β, $\mu_s > l_\beta$.

Small values of l_β are always to be expected when β assumes very high values. At a molar decadic extinction coefficient of $\varepsilon = 10^4\,\mathrm{l\,mol^{-1}\,cm^{-1}}$ and a moderate concentration of $\sim 6\,\mathrm{mol\,l^{-1}}$ in a solid substance we contend with an absorption coefficient $\beta = 6\times10^4\,\mathrm{cm^{-1}}$.

This corresponds to an optical pathlength of $l_\beta = 0.16\,\mu\mathrm{m}$. At a modulation frequency of 100 Hz the thermal diffusion length lies in the region of $1\times10^{-4}\,\mathrm{m}$ to $5\times10^{-5}\,\mathrm{m}$ (100–50 μm), [56] i.e. the requirement for signal saturation $\mu_s > l_\beta$ is met.

Since extinction coefficients of $10^4 \leqslant \varepsilon \leqslant 5\times10^4\,\mathrm{mol^{-1}\,cm^{-1}}$ are very common in many inorganic and particularly in organic compounds this effect must always be expected when investigating solid substances. However, PAS itself provides the possibility of eliminating the saturation effect [57]. Equation (60b) shows that the thermal diffusion length is inversely proportional to the root of the modulation frequency ($\omega = 2\pi f$). If the modulation frequency ω is increased the thermal diffusion length μ_s decreases. By varying ω, l_β can eventually be made to exceed μ_s i.e. we move from case 2a) to limiting case 2b) where, according to Eq. (61), the PA-signal again depends linearly on absorption coefficient β.

In practice when making PA-measurements, the signal amplitude $S_s^{PA} = |\Delta S_s^{PA}|$ of a sample is usually referenced against the signal amplitude $S_r^{RA} = |\Delta S_r^{RA}|$ of a reference sample. The thermal properties and surface reflecting power of this reference sample should be as similar as possible to those of the sample. Therefore, the sample should also have a photoacoustic conversion coefficient $\eta_r = 1$, i.e. all of the absorbed light should be converted into heat without any losses. Furthermore, it must have such a high optical absorbing power that its PA-signal is saturated photoacoustically for the selected modulation frequency, i.e. a PA-signal as shown in Eq. (62) is generated. If S_s^{PA} is rationed against the reference signal S_r^{PA} we obtain the normalised PA-signal amplitude S_n^{PA}:

$$S_n^{PA} = \frac{S_s^{PA}}{S_r^{PA}}\ .$$

From Eq. (61), with $\mu_g = (2\alpha_g/\omega)^{1/2}$, $\mu_s = (2\alpha_s/\omega)^{1/2}$, and $K_s = \alpha_s \varrho_s C_{ps}$ and taking account of B, the signal amplitude S^{PA} is obtained as:

$$S_s^{PA} = K_s I_0(\lambda)\beta_s \mu_s \tag{61a}$$

with

$$K_s = \frac{\gamma P_0 \eta_s}{2 l_g T_0 \omega \varrho_s C_{ps}} \cdot \sqrt{\frac{\alpha_g}{\alpha_s}}\, Q_{aku}^{(\omega)}(\omega) S^M(\omega)$$

and analogously there is obtained from Eq. (62):

$$S_r^{PA} = K_r I_0(\lambda)$$

with

$$K_r = \frac{\lambda P_0 \eta_r}{2 l_g T_0 \omega \varrho_r C_{pr}} \sqrt{\frac{\alpha_g}{\alpha_r}}\, Q_{aku}^{(\omega)} S_{(\omega)}^{M} \ .$$

Thus, for a normalized PA-signal amplitude, the following applies

$$S_n^{PA} = \frac{K_s}{K_r} \beta_s \mu_s = K_n \beta_s \mu_s$$

with

$$K_n = \frac{\varrho_r C_{pr}}{\varrho_s C_{ps}} \sqrt{\frac{\alpha_r}{\alpha_s}}\, \eta_s \ .$$

Here, the same experimental conditions for sample and reference are assumed, i.e. $Q_{aku}(\omega)$ and $S^M(\omega)$ are the same as well as $I_0(\lambda)$, γ, P_0, T_0, ω and l_g. We also presuppose that $\alpha_g(s) \simeq \alpha_g\, \alpha(r)$ which always applies to solid substances and, provided that the same solvent is used for sample and reference, to liquid samples also.

The introduction of a normalized PA-signal amplitude leads by comparison with Eqs. (64) and (65), without any simplifying assumptions, to a clearer picture of the correlation between the PA-signal amplitude S_n and the absorption coefficient β_s.

In principle, the equation for S_n^{PA} permits the determination of β_s if the thermal data for sample and reference included in K_n and the conversion coefficient η_s of the sample are known accurately. However, these requirements are not easily met and the determination of β_s from S_n^{PA} is frequently problematical. As in the case of Eq. (64) or (65), we assume that K_n $(= K_s/K_r)$ is determined with the spectrophotometrically measured β values.

The normalization of the PA-signal amplitude also permits true double-beam measurements.

Figure 38 shows an outline of a double-beam PA-spectrophotometer with a computer control. The optical part with light source (Osram xenon high pressure lamp XBO 450 W/1) and monochromator correspond to the arrangement used for conventional absorption spectroscopy. Permanently fixed microphones such as condenser microphones are substituted for the photomultiplier of the reference (PA_R) and measurement (PA_S) cell.

The design of a double-beam PA-spectrophotometer without computer has been described [58]. A ratiometer, the output of which is coupled directly to a compensating recorder, is substituted for the ADC.

The gas-coupled PA-measurement cell is a Helmholtz resonator which has its highest sensitivity at a specific resonance frequency [59]. In order to utilize the advantages of PA-spectroscopy this resonance frequency should be as high as possible to allow measurements with a variable chopper frequency over a broad region. This may be achieved by making the distance between the microphone membrane and gas volume short. The total volume of the measurement cell should be kept as small as possible ($1-2\ cm^3$).

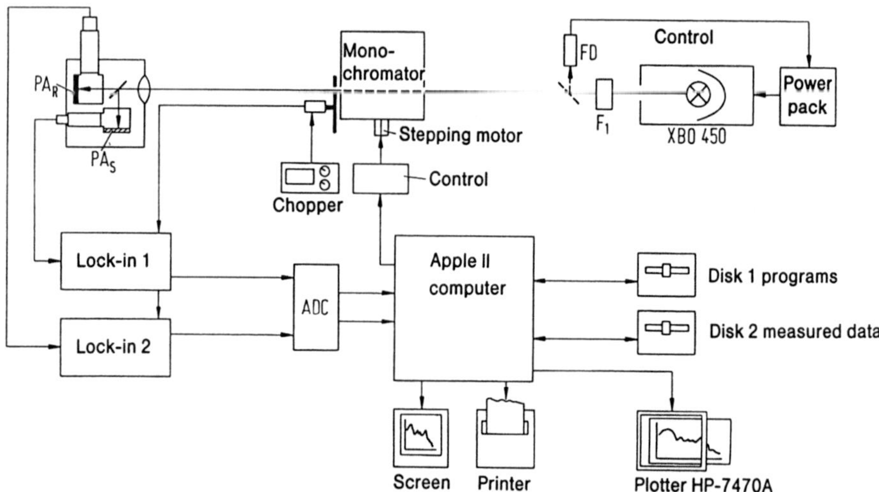

Fig. 38. Design of a photoacoustic dual-beam spectrophotometer with computer control

Cells with these dimensions have resonance frequencies in the region of 2000–4000 Hz. This provides a broad region from ~2 Hz upwards for PA-investigations.

5.4.2 PAS Applications

Rosencwaig has given a comprehensive description of the versatility of PAS applications in physics, technical physics, chemistry, physical and analytical chemistry as well as in biology, biochemistry and medicine [50]. The principles of PAS and its applications to gases, liquids and solid substances have also been discussed in the book "Optoacoustic Spectroscopy and Detection" [61]. The examples show that PAS is a method which can be used in many ways. However, the older works lack a critique of the power of this method particularly when compared with other spectroscopic methods in the UV-VIS and IR region. Generally, PAS should be applied to problems which cannot be solved by conventional spectroscopic methods. PAS should also be employed if it truly complements luminescence and diffuse reflectance spectroscopy.

The RG theory is derived for a homogeneous and isotropic condensed phase. If the validity of this theory is accepted then confirmation of the theoretical relationships can only be expected when homogeneous and isotropic solid substances are under investigation. Amorphous glasses such as filter glasses are an ideal system for PA-investigations. On account of their range of transmission values throughout the near UV, VIS and NIR region they also permit the investigation of the transition from optically transparent to optically opaque samples.

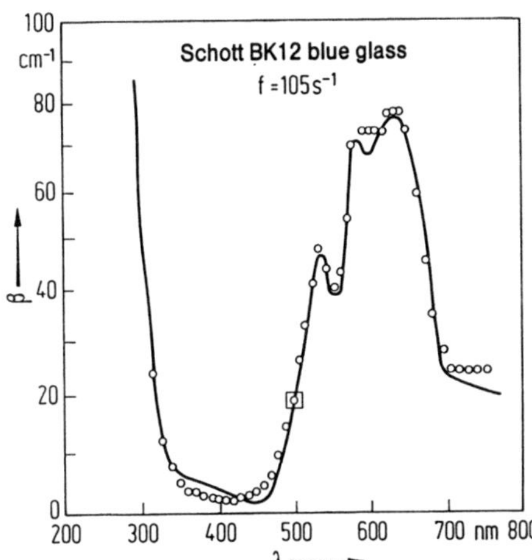

Fig. 39. Photoacoustic spectrum of filter glass BG 12

Figure 39 shows a PA-spectrum of a *thin Schott filter glass* BG 12 at a modulation frequency of 105 s^{-1} [52]. The transmission of this filter glass was spectrophotometrically determined at $\lambda = 500$ nm (reference point: drawn on the curve as a rectangle). The absorption coefficient is $\beta = 16\ cm^{-1}$. If the PA-signal is determined at the same wavelength the constant factor F can be calculated by means of Eq. (64). If the thermal diffusivity α_s is known, the PA-signals measured at any wavelength can be converted into the corresponding β values at the given modulation frequency. These values can then be plotted as ordinate against the wavelength.

The thermal diffusivity of neutral glass NG 1 was used ($\alpha_s = 5.07\ 10^{-3}\ cm^2\ s^{-1}$) [62].

In practice, spectrophotometric measurements can only be made for $\beta < 20\ cm^{-1}$ with a filter with a pathlength > 3 mm. The manufacturer has tabulated a complete transmission curve of this filter for d = 1 mm [63]. The β-values calculated for the measured pathlength are plotted as circles. The example shows that the PA-spectrum calculated using Eq. (64) is in full agreement with the spectrum calculated from transmission measurements (β-values).

A filter glass with β-values > 50 cm^{-1} can be described as opaque which means that PAS is superior to conventional spectroscopy in this case. Further investigations have been carried out with filter glasses NG 1 and RG 1000 [52].

When samples are investigated which are not homogeneous in the direction of the irradiation, then PA-spectra show a significant dependence on the modulation frequency [64, 65]. An example is given in Fig. 40. Colloidal silver has been diffused into the glass from one side and so there is an inhomogeneous distribution of colloidal silver from the surface to the inside of the glass (a concentration profile) [52]. The PA-spectra for modulation frequencies of 15, 105 and 7400 s^{-1} show a significant dependence on the

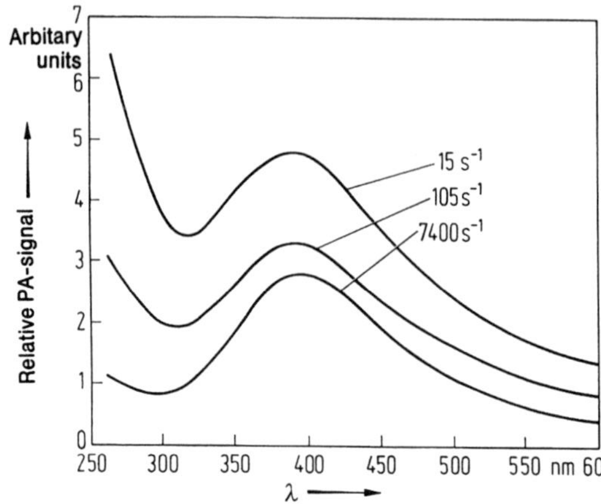

Fig. 40. Photoacoustic spectrum at different modulation frequencies of a glass containing diffused colloidal silver

modulation frequency. The absorption of the glass itself takes place over the whole pathlength of a sample in the region of $\lambda < 300$ nm. However, the thermal diffusion length μ_s is, by Eq. (60b), a function of the modulation frequency. Therefore, the thermal diffusion length μ_s decreases with increasing frequency for these three measurements. Consequently, the absorption of the glass is suppressed in PA-spectra relative to the absorption at the color centers, which can be seen particularly clearly at 7400 s^{-1}. From this it can be concluded that the color centers lie just below the sample surface.

In agreement with this conclusion, the absorption bands of *colloidal silver* are not observed in PAS at $\lambda = 400$ nm if measurements are taken from the other side of the glass. A conventional absorption measurement would be totally unsuitable here since the absorption is summed over the whole pathlength in transmission measurements [52].

This effect of different PA-spectra when measuring from the front and rear is known as the *lateral specifity* of this method. We see that PAS is superior to conventional absorption spectroscopy for such problems since it can supply valuable *additional* information.

Initially, it was thought that the absorbing power of samples was unimportant in PAS. For that reason, PA-investigations were frequently carried out on solid *powdered samples* [60]. It was found that a PA-spectrum with generally excellent signal intensities could be obtained. However, the theoretical interpretation was considerably more difficult since the requirements of the RG theory were no longer satisfied. In such investigations the effect of signal saturation must always be expected since very high absorption coefficients can occur, particularly in organic pigment powders. Such samples behave similarly in reflectance spectroscopy where strongly absorbing samples fail to provide diffuse reflectance spectra, owing to the strong specular component; formally, signal saturation exists over a broad spectral region.

Here a dilution technique can be utilized for obtaining suitable PA-spectra [54, 66, 67, 68].

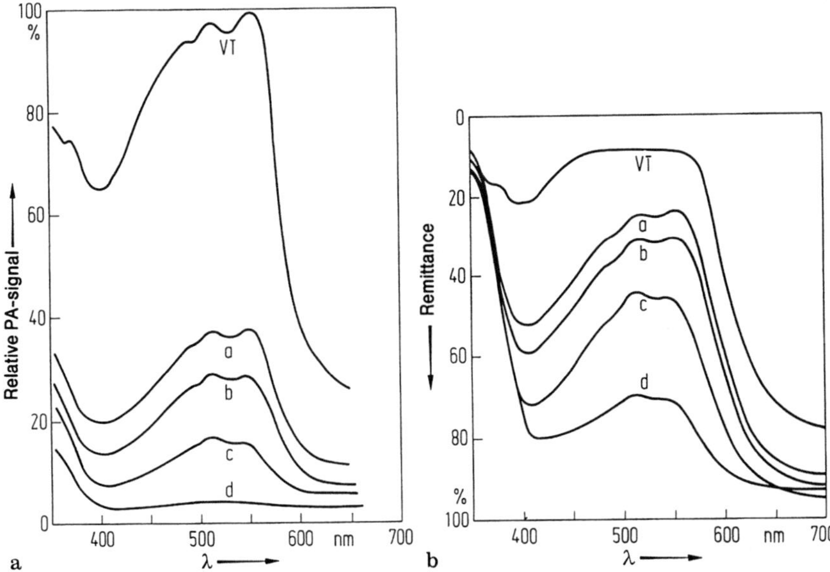

Fig. 41 a, b. Photoacoustic spectra of a quinacridone pigment; modulation frequency 5050 Hz; reflectance spectrum for comparison; serial dilution with TiO_2: VT full tone, *a* 1:40, *b* 1:100, *c* 1:1000, *d* 1:10000

A comparison between reflectance and PA-spectra shows initially that, at *low* modulation frequencies, PAS has no obvious advantage vis-a-vis reflectance spectroscopy. However, the benefit of PAS is seen immediately when higher modulation frequencies $\omega = 2\pi f$ [67] are used. Figure 41 shows the PA- and reflectance spectra of a quinacridone pigment (Color Index violet No. 19) serially diluted with TiO_2. The PA-spectra were measured at a modulation frequency of $5050\,s^{-1}$ [67]. Whilst no information about the absorption spectrum is obtained from a reflectance spectrum of the full tone (undiluted pigment) in Fig. 41 b, the PA-spectrum of the full tone (Fig. 41 a) shows a complete band structure in the region of 400–600 nm.

Tilgner and Lüscher have posed the question as to whether PAS is superior to conventional reflectance spectroscopy for powders [68]. This question has been raised again by Burggraf and Leyden [54] and by Brücher et al. [67]. They compared the PAS parameters with those of the Kubelka-Munk function, i.e. with the scattering and absorption coefficient. As a result of these PA-measurements the possibility of evaluating powdery samples quantitatively was realized (see Meichenin and Auzel [69]). Another advantage of PAS is that, in addition to measuring the amplitude of the PA-signal, its phase, i.e. the phase angle can also be determined [67]. Although the systems outlined briefly here do not meet the requirements of the RG theory, it is to be expected from this theory that a phase-angle spectrum will be considerably less influenced by signal saturation [70].

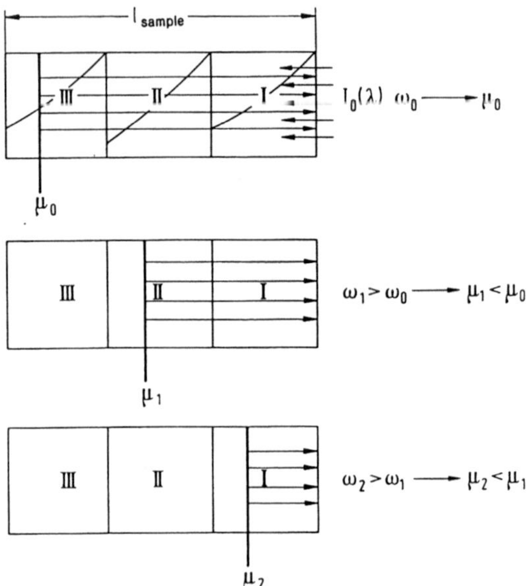

$$\mu = \sqrt{\frac{2\alpha}{\omega}} \; ; \; Q_{PA} \sim \frac{\beta \mu_{sample}}{2a_g} \left(\frac{\mu_{sample}}{K_{sample}} \right) K I_0$$

Fig. 42. Depth-profile determination by means of photoacoustic spectroscopy

As with diffuse reflectance spectroscopy, the dilution technique with a white standard such as TiO_2, MgO, Al_2O_3 etc. gives us the opportunity to investigate species absorbed on these partly acid or basic oxide surfaces by means of PAS [71–74].

PA-investigations on inorganic *insulators* and *semiconductors* have become classical examples of the application of PAS [60].

In the case of insulators, PA-spectra supply direct information about the optical absorption bands in these materials. The benefit of PAS is always apparent when optical transmission measurements fail, as frequently occurs in the case of compact and powdery materials [75–79].

An interesting PAS application is in the *nondestructive* determination of *depth profiles* [80]. Figure 42 gives a qualitative illustration of the method.

For simplicity, we assume that three differently absorbing layers (e.g. three different dyes) are arranged sequentially with sharply defined boundaries. Furthermore, we presuppose that the matrix in which they are dissolved is the same. Consequently, the heat transfer is determined by the thermal properties of the matrix. As a result, the thermal diffusion length μ_s is also determined by Eq. (60b). This length is a function of modulation frequency $\omega = 2\pi f$ and μ_s is very large at low modulation frequencies. Figure 42 shows this as μ_0. The incident light is absorbed by all three layers and generates the PA-signal, when it is converted into heat. In this case, the measured spectrum will be the result of the superposition of the three PA-spectra.

If the modulation frequency is increased to ω_1 when only the heat wave generated in the region between the sample surface and μ_1 is effective and provides the PA-spectrum. The contribution of absorbing layer III shown in Fig. 42 can no longer be observed. Thus, the PA-spectrum corresponds essentially to the superposition of the spectra from layers II and I. At an even higher modulation frequency, ω_2, we observe only that contribution to the PA-signal generated in layer I, because $\mu_2 < \mu_1$.

If the measurements at ω_0, ω_1 and ω_2 are combined, an image of the arrangement of these three layers, from the inside to the outside, is obtained, i.e. a *depth profile.* In principle, we have already used this technique in Fig. 40.

Examples of this type of application are investigations of intact *plant leaves* [78, 80] and petals [81] whose upper surface usually consists of a waxy layer which only absorbs in the near UV region. Typical three-layer systems are *color reversal papers* and films. Görtz et al. have reported experiments on Agfacolor reversal paper [82]. Parallel studies have been carried out on reversal films [83].

The measurement of the phase angle is also important in these investigations. In the transition between one layer and another the optical and/or thermal properties of a sample change, which, in accordance with the theory, leads to a sudden change of the phase angle [82].

PA-investigations of *solutions* are particularly well suited to testing the theoretical concepts relating to PAS. Aqueous dye solutions have been used in this case [84, 85, 86]. The optical properties of such samples can be conveniently set by an appropriate dilution and if the thermal properties of the solvent are known the PA-signal can be calculated.

If the thermal diffusivity of *pure water* is taken as a basis we obtain $\mu_s = 7 \times 10^{-3}$ at a modulation frequency of $10\,s^{-1}$ from Eq. (60b). Thus, aqueous *dye solutions* can be regarded as thermally thick at the usual pathlengths. Schneider and Coufal [87–90] have reported interesting results of PA-investigations of aqueous and alcoholic dye solutions. They studied the question of whether an isosoptoacoustic point is observed in the PA-spectra of a dissociation equilibrium analogous to an isosbestic point in the absorption spectra by making use of the pH-dependence of the absorption spectrum of bromothymol blue. Such analogous behaviour is not to be expected a priori because of the fundamental difference between PA- and absorption spectroscopy. However, in the case of the pH-dependence of bromothymol blue, we find excellent agreement between the position of the isosbestic point at $\lambda = 500$ nm and the isosoptoacoustic point [88].

These findings are understandable if it is assumed that both species (protonated and nonprotonated bromothymol blue) present in the protolytic equilibrium possess the same quantum yield for nonradiative deactivation.

Substances in which the absorbed photon energy is converted completely into heat are ideal cases for PAS. However, if two species, whose fluorescence quantum yields differ extensively are present in a solution, it is to be expected that an isosoptoacoustic point will not be observed. A *mix-*

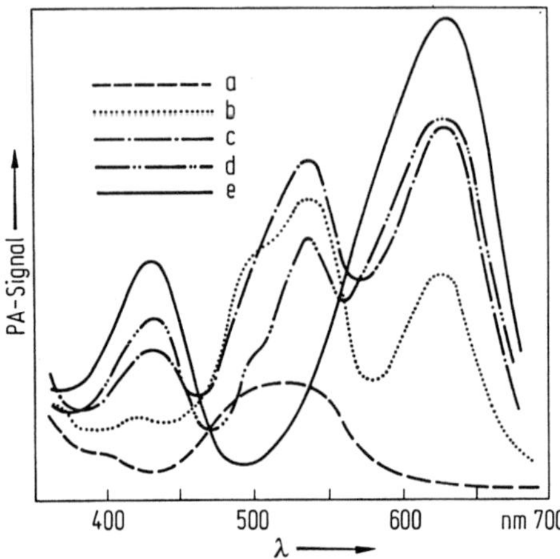

Fig. 43. Photoacoustic spectra of rhodamine/malachite green solutions. *a* 5:0 only rhodamine 6G; *b* 4:1 rhodamine 6G/malachite green; *c* 2.5:2.5 rhodamine 6G/malachite green, *d* 1:4 rhodamine 6G/malachite green; *e* 0:5 only malachite green

ture of rhodamine 6G and malachite green is an example of such a model system.

(Φ_{FM}(rhodamine 6G) = 0.8; Φ_{FM}(malachite green) $\leq$ 0.1).

Figure 43 shows the PA-spectra of such mixtures taken from the measurements of Schneider and Coufal [87, 88]. Pure solutions of rhodamine 6G (curve a) and malachite green (curve e) show very different PA-spectra. Whilst the PA-spectrum of malachite green approximates to an absorption spectrum, an equimolar solution of rhodamine 6G exhibits a typical saturation effect and overall a lower signal intensity which can be traced back to the radiative deactivation process which dominates here.

Since malachite green does not absorb at $\lambda = 500$ nm (see curve e) a definite signal increase in the region of the absorption maximum of rhodamine 6G can be observed for mixtures. This increase displays a maximum value at a mixture ratio of 1 : 1 (curve c). Whilst optical measurements on these systems show several isosbestic points no isosoptoacoustic point can be recognized in Fig. 43.

This behaviour can be explained by an energy transfer occurring in the system between rhodamine 6G as donor and malachite green as acceptor.

A practical application of energy transfer is the targeted addition of a nonfluorescent substance making it possible to detect molecules with a small photoacoustic signal yield (large fluorescence quantum yield).

With the help of PA-phase angle spectroscopy, absorption coefficients such as those of aqueous dye solutions can also be determined. For the case described by Eq. (61), a correlation between the normalised phase angle Φ and the absorption coefficient β_s is given by the RG theory in the form of the Poulet, Chambron, Unterreiner (PCU) approximation [91, 92]:

$$\Phi = -(\pi/2)) \text{ arc tan } (1+\beta_s\mu_s) \ . \tag{68}$$

From the inverse of Eq. (68) we obtain:

$$\beta_s = [\tan(\Phi + \pi/2) - 1]/\mu_s \ . \tag{68a}$$

Phase-angle spectroscopy is a real alternative to conventional absorption spectroscopy. This applies particularly to concentrated solutions where spectrophotometric measurements of β_s are often not possible on account of the short pathlengths required.

Thermoelastic oscillations in the solution cannot be neglected in dilute solutions. However, they are considered in the CP model [51, 53]. The phase-angle function $\Phi = F(\beta_s)$ given by Wetsel [51] does not accurately reproduce the experimentally determined correlation between phase angle Φ and absorption coefficient β_s. In contrast, a semiempirical phase-angle function [93] gives the form of the function $\Phi = F(\beta)$ in the region of $0 \leq \beta_s \leq \infty$ in good agreement with experiment:

$$\Phi = -\pi/2 + \arctan[1 + \beta\mu_s + P_z/\beta^{P_e}] \ . \tag{69}$$

P_z and P_e are empirical parameters of a general hyperbolic function which allows for the contribution of thermoelastic oscillations and takes better account of the observed phase-angle behaviour, $\lim_{\beta \to 0} \Phi = 0$, than the extended CP model [51].

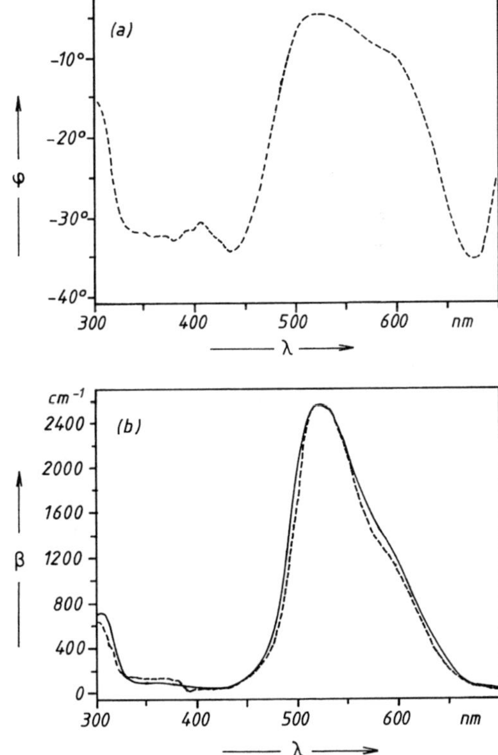

Fig. 44a. Experimental values of the photoacoustic phase angle spectra of Crystalviolet dissolved in water **b** absorption spectra derived from **a** (......) and compared with the absorption spectra measured by photometric methods (———). Concentration: 1.09×10^{-2} mol/l; Modulation frequency f = 28 cyles/s

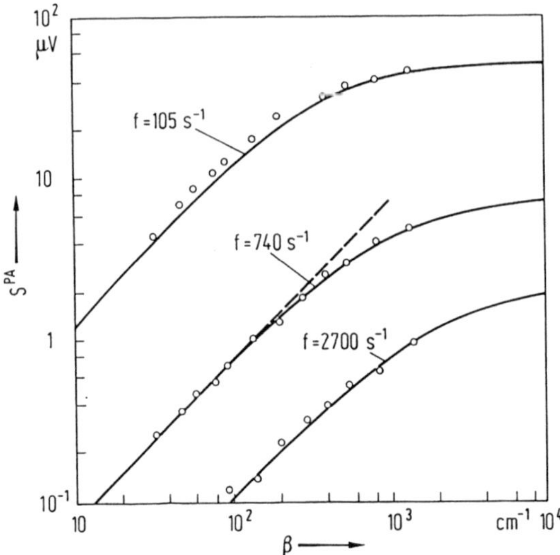

Fig. 45. Photoacoustic signals of phenol red in polyvinyl pyrrolidine versus the absorption coefficient at three different modulation frequencies. Continuous curves calculated from the RG theory

Equation (69) cannot be inverted. However, complete absorption spectra of synchronously measured PA-amplitudes and PA-phase angle values, which show excellent agreement with spectrophotometrically measured absorption spectra, can be calculated by means of an iteration technique [94]. In Fig. 44, the absorption spectrum of an aqueous crystal violet solution ($c = 1.09 \times 10^{-2}$ mol/l), calculated according to Eq. (69), is compared with a spectrophotometrically measured absorption spectrum.

Some of the *difficulties and sources of error* in PA-investigations of liquid solutions with gas-coupled measurement cells can be excluded by using *solid solutions* in polymer matrices [68]. Although, in principle, molecularly dispersed solutions are involved, the measurements can be evaluated as for those on solid samples. Görtz and Perkampus have shown that such systems are ideal model systems for testing the RG theory [95]. If solid polymer films or foils are thermally thick they can always be regarded as cases 1 and 2b) of Eq. (61), i.e. the requirements of the RG theory are met under these conditions [95].

These solid solutions have another advantage because the absorption coefficient β or extinction A can be varied directly within broad limits if the thermal properties of the polymer matrix are constant. Thus, the whole range from optically transparent to optically opaque samples can be conveniently covered. Of course, this applies to the same extent to solutions. Figure 45 shows the dependence of PA-measurements for *phenol red dye in PVP* (polyvinyl pyrrolidine) on the absorption coefficient β for three different modulation frequencies. We see that a linear relationship as shown in Eq. (61) can be observed for all three modulation frequencies in the region of small optical densities. This gives way to saturation at higher β-values, see Eq. (62). In agreement with the RG theory, signal saturation sets in at higher optical densities for higher modulation frequencies than at $f = 105\ s^{-1}$. The RG theory, in the form of the PCU approximation, provides a more accurate representation of the curve shown in Fig. 45.

In previous examples it was variously assumed that a PA-signal for a specific modulation frequency can be calculated by Eq. (61) or (62) or by the complete RG theory [48, 49]. However, this always presupposes that we know the thermal data and absorption coefficients of the samples.

If the thermal data are unknown, one can readily see from Eqs. (61) and (62) that the thermal data of the sample, in this case the polymer matrix or solvent, can be determined from the measured PA-signal for given β-values at a fixed modulation frequency.

However, the samples must be thermally thick [95].

We pointed out at the beginning that *fluorescence* is a process which competes with the PA-effect. Other processes to be considered in specific systems were initially excluded.

According to Eq. (59) the following applies:

$$\Phi_{GM} = 1 - \Phi_{FM} \ . \qquad (59)$$

This equation shows a simple relationship between the quantum yield of the nonradiative process Φ_{GM} and fluorescence Φ_{FM}. If we know Φ_{GM}, Φ_{FM} becomes directly accessible.

If S^{GM}, is the photoacoustic signal of a nonfluorescent sample and S^{FM}, the corresponding signal of a fluorescent sample under the same conditions, we obtain the fluorescence quantum yield as:

$$\Phi_{FM} = \frac{\tilde{\nu}_n}{\tilde{\nu}_f}\left(1 - \frac{S^{FM}}{S^{GM}}\right) \ . \qquad (70)$$

($\tilde{\nu}_n$: excitation wavenumber; $\tilde{\nu}_f$: wavenumber of the center of gravity of the fluorescence)

The fluorescence quantum yields can be determined by means of Eq. (70) [97–102].

As mentioned above, a PA-measuring cell has its maximum sensitivity at its resonance frequency [59]. If the resonance frequency is selected as the working frequency we speak of *resonance PAS*. Hess has investigated this in detail [103].

The *temperature* is another parameter determining the magnitude of the PA-signal in a gas-coupled measuring cell. In factor B, see Eq. (61), the temperature is in the denominator. Consequently, the intensity of a PA-signal increases considerably if we lower the temperature of a sample. For this reason, several *low-temperature PA cells*, which can be cooled to the temperature of liquid nitrogen or liquid helium, have been designed [104–106].

Bechthold has developed a PA cell for *high temperatures* (300–1050 K region) [107].

The *piezoelectric detector* has proved itself as a photoacoustic detector. Breuer has described its theory and applications [108]. Tam and Patel have supplemented the operating technique [109].

Hey and Gollnick have described an application of the photoacoustic effect in the form of a *relaxation method* [110].

5.5 Luminescence-Excitation Spectroscopy

In luminescence-excitation spectroscopy we utilize the *radiative* deactivation processes, fluorescence or phosphorescence after excitation of a molecule, i.e. we measure fluorescence (FE) or phosphorescence (PhE) spectra. In this technique the wavenumber or wavelength of the fluorescence maximum $\tilde{\nu}_F$, λ_F is set and the exciting wavenumber or wavelength is varied over the absorption spectrum of a sample. The fluorescence intensity as a function of the excitation wavelength reflects the absorption spectrum of the compound concerned. However, the energy distribution of lamp plus monochromator must be taken into account.

The fact that we must again consider the competition between radiative and nonradiative processes means that the fluorescence intensity is not always simply proportional to the absorbed light intensity according to the Bouguer-Lambert-Beer law. A general expression concerning the fluorescence intensity of an isotropically distributed quantity of emitting species k in a low-viscosity solution is given in reference [111]:

$$F_k(\lambda_i, \lambda_j) = K \cdot q_k(\lambda_i, \lambda_j) \frac{2{,}303\, \varepsilon_k(\lambda_i) \cdot c_k}{S_k(\lambda_i)} \cdot I_0(\lambda_i) \{1 - \exp[-S_k(\lambda_i) d]\} \tag{71}$$

where

$F_k(\lambda_i, \lambda_j)$	is the fluorescence intensity in Einsteins after excitation at wavelength λ_i at bandwidth $\Delta\lambda_i$,
$q_k(\lambda_i, \lambda_j)$	is the spectral quantum yield of component k,
$\varepsilon_k(\lambda_i)$	is the molar decadic extinction coefficient of component k ($m^2\,mol^{-1}$),
c_k	is the concentration of component k ($mol\,m^{-3}$),
$S_k(\lambda_i) = 2.303 \sum_{k=1}^{r} \varepsilon_k(\lambda_i) c_k$	with $\varepsilon_k(\lambda_i)$ the molar extinction coefficient at wavelength λ_i and c_k, the molar concentration of the k'th component,
$I_0(\lambda_i)$	is the irradiating light intensity in Einsteins at wavelength λ_i,
d	is the pathlength,
λ_i	is the wavelength of irradiation,
λ_j	is the observation wavelength (λ_F).

The quantity K is a constant of the apparatus which is independent of the wavelength in the formulation given by Eq. (71), i.e. wavelength for grating monochromators and a fixed slit width. When formulating Eq. (71) in wavenumbers, K changes as λ^{-2}.

If the exponential function in Eq. (71) is expanded in a series it can be terminated after the second term for dilute solutions since c_k is very small.

$$F_i(\lambda_i, \lambda_j) = I_0(\lambda_i) \cdot 2.303\, q_k(\lambda_i, \lambda_j) \cdot \varepsilon_k(\lambda_i) \cdot c_k \cdot d \cdot K \; . \tag{72}$$

The fluorescence intensity is directly proportional to the extinction coefficient $\varepsilon_k(\lambda_i)$ in this relationship and the connection with the absorption spectrum can be seen immediately.

Furthermore, the fluorescence intensity is proportional to concentration c_k in Eq. (72) which thus forms the basis for quantitative fluorimetry [112–114]. However, the value of the product $2.303 \times \varepsilon_k \times c_k \times d$ must be smaller than 0.1 [111]. If the exponent is very large (e.g. at high concentrations and high extinction coefficients) the expression $\exp\{-S_k(\lambda_i)d\}$ approaches zero and all light is completely absorbed within a low penetration depth. Fluorescence light is now generated in a narrow layer of the solution. The excitation spectrum cannot be measured under these conditions and evaluation is not possible. The reabsorption of the fluorescence is also to be expected in concentrated solutions. This is particularly troublesome if the 0–0 transitions in absorption and fluorescence are only slightly shifted from each other.

In addition to the essential requirement of a *dilute* solution, another requirement must be met, the fluorescence or phosphorescence must always be emitted from the lowest excited singlet state, S_1, or triplet state, T_1. Apart from a few exceptions, this is always the case in accord with Kasha's rule [115].

Figure 46 shows the layout of a luminescence spectrophotometer. The instrument can be used for the determination of fluorescence, phosphorescence and excitation spectra [112, 114].

For excitation spectra, monochromator b_2 (λ_j) is set at the maximum of the luminescence spectrum and the excitation wavelength λ_i is continuously varied over the absorption spectrum of the sample by monochromator b_1. In order to obtain a quantum proportional (corrected) luminescence spectrum, part of the exciting light can be diverted onto a quantum counter via the beam splitter d. The luminescence signal from the amplifier is divided in the ratiometer k by the signal from the quantum counter and the result is recorded as a function of excitation wavelength λ_i.

In order to keep the reabsorption of fluorescence light as small as possible, the excitation light is directed at a 30° angle into the front window of the cuvette.

Modern fluorimeters usually enable intensity corrected luminescence spectra to be recorded directly (see, for example, the luminescence spectrometer LS-5 from Perkin Elmer).

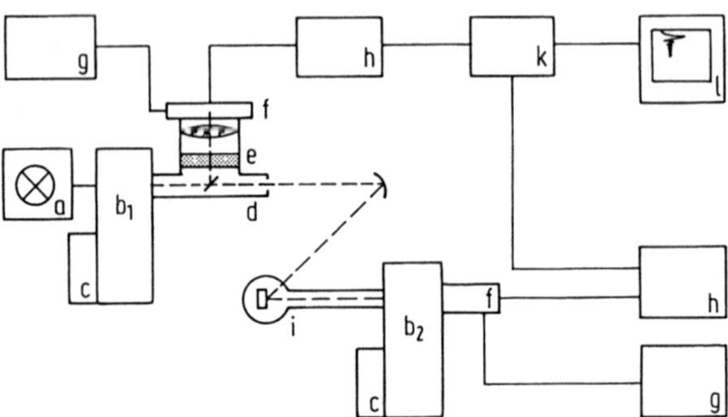

Fig. 46. Outline of a fluorescence spectrophotometer for measuring excitation and luminescence spectra. *a* light source, *b* monochromator, *c* wavelength advance, *d* beam splitter, *e* quantum counter, *f* photomultiplier, *g* high voltage supply, *h* amplifier, *i* cuvette compartment, *k* ratiometer (divider), *l* compensating recorder

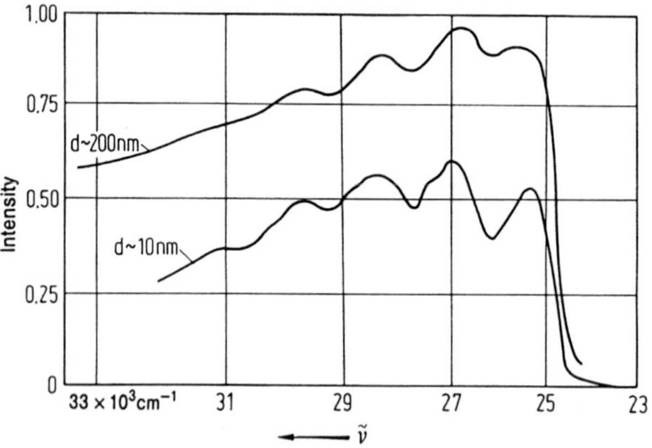

Fig. 47. Fluorescence excitation spectra of two thin anthracene films of differing thickness

Luminescence-excitation spectroscopy is very important for *molecular spectroscopy*. The solutions to be measured must be dilute (see above). In the case of solid substances such as thin films on quartz supports the pathlength must be taken into consideration. Figure 47 shows the example of two *anthracene films* of different thickness. Whilst the film with a pathlength of about 10 nm clearly shows the vibrational structure of the long wavelength absorption bands, we observe considerable flattening at 200 nm pathlength [116]. At even longer pathlengths, a fluorescence intensity corresponding to the spectral energy distribution of the light source would in practice be recorded over the absorption wavenumber range. The pathlength dependence in *photoaction spectra* must be taken into account in an analogous manner. In these spectra, the conductivity, instead of the fluorescence intensity, is used as an indicator of the absorbed light energy [117]. This spectroscopic method plays an important role when investigating solids and in particular solid organic substances. By means of FE spectroscopy changes in a system during exposure can also be determined.

The excitation spectrum of *delayed fluorescence* can also be measured by means of tunable *dye lasers*, due to their high intensities. Delayed fluorescence is caused by the mutual annihilation of *two* triplet excitons [118] which generates a singlet exciton which can return radiatively to the ground state:

$$S_0 + h\nu \rightarrow T_1$$

$$T_1 + T_1 \rightarrow S_1 + S_0$$

$$S_1 \rightarrow S_0 + h\nu_F \ .$$

Experimentally, we verify the participation of two triplet excitons in generating delayed fluorescence by observing that the delayed fluorescence signal increases with the square of the exciting-light intensity [119, 120].

If the excitation wavenumber is changed continuously in the region of the triplet state, a corresponding dependence of the triplet exciton concentrations is obtained from the wavenumber dependence of the extinction coefficients. Thus, the signal intensity of delayed fluorescence also becomes a function of the excitation wavenumber. In the delayed fluorescence excitation spectrum the fluorescence signal is plotted against excitation wavenumber ($\tilde{\nu}_i$) or excitation wavelength (λ_i) to obtain the absorption spectrum of the $S_0 \rightarrow T_1$ transition [119, 120].

Another variant concerns the *use of polarized light.* This leads to "photoselection" which is extremely important in molecular-physical problems [121, 122]. The basis is the fact that, by Kasha's rule, fluorescence always arises from the lowest excited state, regardless of the electronic excited state to which a molecule is exited. However, anisotropy of light absorption (see Sect. 2.4) is present in many unsaturated molecules and particularly in planar ones. These effects are characteristic of condensed aromatic hydrocarbons and their heteroanalogs.

We take a fixed *anthracene molecule* of symmetry D_{2h} as an example. The long-wavelength transition, ^1L_a, is polarized along the short molecular axis and the second electronic transition, ^1B_b, along the perpendicular long molecular axis. By irradiating with polarized light whose polarization plane lies parallel to the short axis of the anthracene molecule, the ^1L_a transition polarized along this axis is excited. In order to excite the ^1B_b transition we would have to rotate the plane of the polarized light at the appropriate wavenumber by 90°. If we vary the wavelength of the polarized light over the absorption spectrum then only those molecules whose axes lie parallel to the plane of the polarized light are excited. Since fluorescence always results from the lowest singlet excitated state, if the second excited state is irradiated, the plane of the polarized fluorescence is rotated by 90° because the transition moments are orthogonal. In order to observe this effect we must ensure that the molecules do not change their position in space during the lifetime of the lowest singlet excited state ($10^{-9}-10^{-7}$ s).

Thus, we work with glassy solidified solutions at the temperature of liquid nitrogen or with polymer films. The polarized light selects only those molecules from the statistically distributed mass whose transition moments lie in the plane of the polarized light at the appropriate wavelength. Therefore this method is called the *photoselection method.*

We can distinguish between the different ways of carrying out the experiment and the corresponding spectra:

a) We measure the degree of polarization the fluorescence at a fixed wavelength λ_j with a variable excitation wavelength λ_i. This is the *a*bsorption-*p*olarization-*f*luorescence spectrum (APF spectrum).
b) We measure the degree of polarization over the fluorescence spectrum at a fixed excitation wavelength λ_i. This is the *f*luorescence-*p*olarization spectrum (FP spectrum).
c) By analogy with a), we obtain the *a*bsorption-*p*olarization-*ph*osphorence spectrum (APPh spectrum) from the phosphorescence.
d) By analogy with b), we obtain the *ph*osphorescence-*p*olarization spectrum (PhP spectrum).

The transition from one electronic excited state to the next higher one is marked by a change of polarization in the APF spectrum from positive to negative values. Figure 48 shows as an example the APF spectrum of

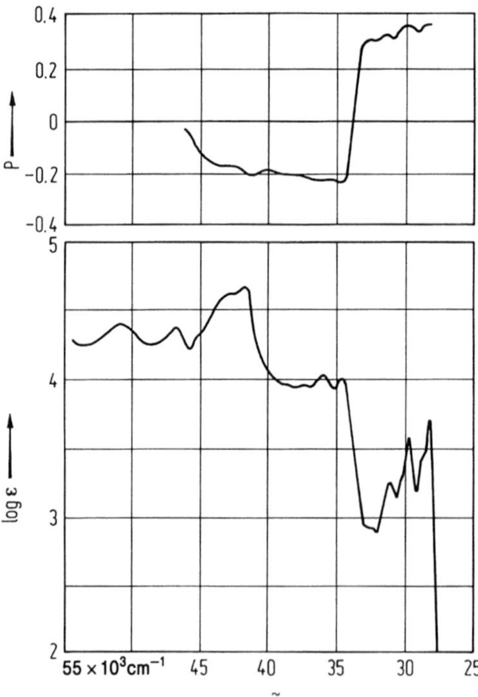

Fig. 48. Absorption-polarization-fluorescence spectra of 2,7-diazaphenanthrene; solution 77 K

2,7-diazaphenanthrene which clearly exhibits the change of sign of polarization in the APF spectrum at ca. 34000 cm^{-1}. This permits an accurate assignment of the electronic transitions.

The polarization, plotted as a function of wavenumber or wavelength, is given by:

$$P = \frac{I_{\parallel} - I_{\perp}}{I_{\parallel} + I_{\perp}} \quad \text{or} \quad P = \frac{3\cos^2\alpha - 1}{\cos^2\alpha + 3} \tag{73}$$

where α is the angle between the transition moments for the absorption and the emission process. For $\alpha = 0$ (parallel) we obtain P = 0.5 and for $\alpha = 90°$ (vertical) we obtain P = −0.33 [122]. This applies to all molecules with twofold axes of rotation.

Figure 48 shows that APF spectra are usually structured which can be traced back to the coupling between electronic and vibrational excitation.

Dörr should be consulted for details of the experimental techniques and other particulars [118, 122].

In the *analytical application* of FE spectroscopy it is frequently necessary to detect in a sample species whose fluorescence spectra more or less overlap. If the components have different absorption spectra it is possible to excite one or more components separately by varying the excitation wavelength. In this way, a simple, and possibly known, fluorescence spectrum may be observed in some circumstances. Parker [124] has described this technique of *selective fluorescence excitation* in detail. The method can also be applied to phosphorescence spectroscopy where the detection of contaminants is particularly important.

An advance on selective fluorescence excitation is the *synchronous fluorescence-excitation spectroscopy* developed by Lloyd [125] who has used it to identify aromatic hydrocarbons, petroleum products and crude oils [126]. The excitation wavelength λ_i and observation wavelength λ_j are varied synchronously such that a constant wavelength difference $\Delta\lambda$ exists between the two wavelengths. The spectra recorded over λ_j depend strongly on the magnitude of $\Delta\lambda$ for which, for example, 20, 30 or 40 nm can be selected.

The principle of this method is as follows: The fluorescence intensity of a mixture of different components is both a function of the excitation wavelength λ_i and the observation wavelength λ_j. In standard fluorescence spectroscopy, λ_i is kept constant and the fluorescence spectrum is recorded as a function of λ_j. However, with several components in a solution it can no longer be guaranteed that the excitation wavelength λ_i corresponds to the absorption maxima of the components. Thus, there is considerable influence upon the contribution of the individual components to the total fluorescence.

In synchronous fluorescence-excitation spectroscopy the variation of excitation wavelength λ_i is linked to that of observation wavelength λ_j. This has the advantage that the absorption maximum is automatically selected as the excitation wavelength for each component. Since the maximum of a fluorescence spectrum is displaced toward the red by ca. 10–20 nm vis-a-vis the absorption maximum, we can correlate the $\Delta\lambda < 20$ nm absorption maxima of specific compounds with the associated fluorescence maxima. This method is very applicable to aromatic hydrocarbons which have structured fluorescence spectra at ambient temperatures and where, in dilute solutions, the 0–0 transitions are the most intense in both absorption and emission.

However, two factors must be noted:

1. As mentioned above, at higher concentrations and longer pathlengths, reabsorption increases. Shortening the pathlength and directing the exciting light at less than an angle of 30° may help in this case (see above).
2. Energy transfer can occur in mixtures of aromatic hydrocarbons. Consequently, the fluorescence of one component is lost completely and the fluorescence of another component appears [127]. However, this fluorescence cannot be recorded at $\Delta\lambda \sim 20$ nm since the difference between the absorption maximum of one component and the fluorescence of the other component is greater than 20 nm.

Nevertheless, if both factors are taken into account this technique is a suitable method for the qualitative analysis of *aromatic hydrocarbons in crude oil* [126].

Warner, Christian, Davidson and Callis [128] have given a detailed theoretical description of *multicomponent analysis*. They utilized the fact that the fluorescence intensity of a multicomponent system is a function of both excitation wavelength λ_i and observation wavelength λ_j; and they formulated an *emission-excitation matrix* which can be determined experimentally. Elements M_{ij} of the matrix M represent the fluorescence intensity measured at wavelength λ_j during excitation with wavelength λ_i. If both wavelengths are offset appropriately, one row of the matrix M provides the fluorescence spectrum measured at the individual excitation wavelength λ_i whilst one column of the matrix M shows the fluorescence excitation spectrum of the emission at the specific wavelength λ_j. The information contained in this experimentally determined matrix far exceeds that of synchronous fluorescence excitation spectroscopy.

In a dilute solution, by analogy with Eq. (72) for one component, each element of M_{ij} is given by:

$$M_{ij} = 2.303\,\Phi_f \cdot I_0(\lambda_i) \cdot \varepsilon(\lambda_i) \cdot c \cdot d \cdot K(\lambda_j) \cdot \gamma(\lambda_j) \ . \qquad (74)$$

By comparison with Eq. (72), $K(\lambda_j)$ takes account of the wavelength dependence of the sensitivity of the fluorimeter and $\gamma(\lambda_j)$ of the fraction of fluorescence photons emitted at wavelength λ_j. Φ_f is the total fluorescence quantum yield and thus, a quantity specific to the substance. We can summarize the parameters in Eq. (74) as follows:

$\alpha = 2.303\,\Phi_f c d$ is a quantity specific to the substance
$x_i = I_0(\lambda_i)\,\varepsilon(\lambda_i)$ is the excitation spectrum and

$y_j = \gamma(\lambda_j)K(\lambda_j)$ is a quantity determining the fluorescence spectrum as a function of the wavelength

Thus we write in Eq. (74) $M_{ij} = \alpha x_i y_j$. The matrix M is then given for an individual component by

$$M = \alpha \cdot x \cdot y \ . \qquad (75)$$

If a sample contains several different components r we obtain

$$M = \sum_{k=1}^{r} \alpha_k \cdot \chi_i^k \cdot y_j^k \ . \qquad (76)$$

When experimental data are represented in this way, it is assumed that the optical density at all excitation wavelengths λ_i is considerably smaller than 1 and that the solution is diluted to such an extent that energy transfer between the different components is practically impossible.

The analysis of the data now consists of finding the values r, α_k, x^k and y^k of the matrix M of the observed quantities.

The method permits the detection of the number of independent components which contribute to the total fluorescence and, in a two-component system, to the *fluorescence excitation spectrum* and *fluorescence spectrum* of each component. See [128] for an accurate mathematical discussion of this problem taking account of possible measurement errors. The ambiguity of interpretation in this method lies in the overlapping absorption and fluorescence spectra of the components. This procedure has been discussed in detail for several binary combinations of aromatic hydrocarbons such as anthracene, pyrene, perylene, chrysene and fluoranthrene and the two-component system of octaethylporphine and zinc octaethylporphine [128].

An interesting version of FE-spectroscopy involves excitation with dye lasers. This method has proved itself for the investigation of inorganic solids doped with *lanthanide ions*. Selective luminescence excitation with tunable dye lasers allows fluorescence and excitation spectra of ions present in a single type of a crystallographic environment to be obtained. Ions of rare-earth metals are used here as selectively excitable probes. The method is called the *selectively excited probe ion luminescence method* (SEPIL) [129, 130]. It is suitable for high sensitivity trace element analysis. In [129] a limit of detection of the triply positive erbium ion of 25 fg/ml (25×10^{-15} g/ml) has been reported. Of course, the method assumes that excitable fluorescence levels of the ions to be analyzed lie within the range of a tunable dye laser.

Spectroscopy with polarised light, see also [131].

References

1. Shibata S (1976) Angew Chem 88:150
2. DMS UV-Atlas (1971) Perkampus H-H, Sandemann I, Timmons CJ (Hrsg) Butterworth, London; Verlag Chemie, Weinheim, Vol V, Spectra K1/8 u. K1/9
3. Schmitt A (1979) Labor-Praxis Heft 9
4. Chance B (1951) Rev Sci Instrum 22:634
5. Chance B (1954) Science 120:767
6. Togo T, Yoshida I, Kobayashi H, Neno K (1976) 37. Symp Analyt Chem (Japan Soc Analyt Chem) Preprint S 97
7. Shibata S, Furukawa M, Goto K (1971) XVI. Colloq Spectrosc Internat, Heidelberg, Preprint Vol I, 114
8. Honkawa I (1975) Anal Chim Acta 78:487
9. Honkawa I (1975) Anal Lett 8:901
10. Hammond VJ, Price WC (1953) J Opt Soc Am 43:924
11. Morrison JD (1953) J Chem Phys 21:1767
12. Giese AT, Freude CS (1955) Appl Spectrosc 9:78
13. Bonfiglioli G, Brovetho P (1964) Appl Opt 3:1417
Bonfiglioli G, Brovetho P, Busca G, Levialdi S, Palmieri G, Wanka E (1967) Appl Opt 6:447
14. Green GL, O'Haver TC (1974) Anal Chem 46:2191
15. Talsky G (1981) Techn Messen 48:211
16. Talsky G, Dostal J, Haubensack O (1982) Fresenius Z Anal Chem 311:446
17. Talsky G (1982) GIT, Fachz Lab 26:913
18. Talsky G, Mayring L, Kreuzer H (1978) Angew Chem 90:563
19. Talsky G, Mayring L (1978) Fresenius Z Anal Chem 292:233
20. Juffernbruch J (1982) Dissert, Univers Düsseldorf
21. Schmitt A (1977) Perkin-Elmer, Angew UV-Spektr Heft 1 und 3
22. O'Haver TC, Green GL (1976) Anal Chem 48:312
23. Juffernbruch J, Perkampus H-H (1983) Spectrochim Acta A 39A:905
24. Talsky G (1979) Makromol Chem 180:513
25. Talsky G, Dostal J, Glasbrenner M, Götz-Maler S (1982) Angew Makromol Chem 105:49
26. Shibata S, Furukawa M, Honkawa T (1976) Anal Chim Acta 81:206
27. Shibata S, Furukawa M, Goto K (1973) Anal Chim Acta 65:49
28. Talsky G (1983) Intern J Environ Anal Chem 14:81
29. Cahill JE, Padera FG (1980) Perkin-Elmer Appl Data Bull ADS-122
30. Botton D, Honkawa T, Tohyama S (1977) ibid. ADS-104
31. Schmitt A (1977) Z Clin Chem, Clin Biochem 15:303
32. Kubelka P, Munk F (1931) Z Techn Phys 12:593
33. Kortüm G (1966) Reflexionsspectroskopie. Springer, Berlin Göttingen Heidelberg New York
34. Kortüm G, Kortüm-Seiler M (1947) Z Naturforsch 2a:652
35. Goebel DG (1966) J Opt Soc Am 56:783
36. Kortüm G, Braun W, Herzog G (1963) Angew Chem 75:653; Angew Chem intern Edit 2:333
37. Stenius AS (1951) Svens Pappenstudning 54:663
38. Kortüm G, Schreyer G (1955) Angew Chem 67:694
39. Schwuttke G (1953) Z Angew Phys 5:303
40. Frei RW, McNeil JD (1973) Diffuse Reflectance Spectroscopy in Environmental Problem-Solving. CRC-Press, Cleveland, Ohio
41. Kortüm G, Friz M (1969) Ber Bunsenges Phys Chem 73:605
42. Hezel U (1973) Angew Chem 85:334
43. Hezel U (1977) GIT, Fachz Lab 21:694
44. Zeiss, Chromatogramm-Spektralphotometer, Literaturverz 1977; A 50-675/K18-d
45. Bell AG (1880) Amer J Sci 20:305; Phil Mag 11:510 (1881)

46. Tyndall J (1881) Proc Roy Soc 31:307
47. Röntgen WC (1881) Phil Mag 11:308
48. Rosenczwaig A, Gersho A (1975) Science 190:556
49. Rosenczwaig A, Gersho A (1976) J Appl Chem 47:64
50. Rosenczwaig A (1980) Photoacoustics and Photoacoustic Spectroscopy. Chemical Analysis, Vol 57. (Elring PJ, Wineforder JD (eds)). Wiley, New York
51. Wetsel GC, McDonald FA (1977) Appl Phys Lett 30:252
52. Görtz W, Perkampus H-H (1983) Z Phys Chem NF 134:31; Clarenbach B, Perkampus H-H (1985) Fresenius Z Anal Chem
53. McDonald FA, Wetsel GC (1978) J Appl Phys 49:2313
54. Burggraf LW, Leyden DE (1980) Appl Phys 51:4985
55. Pelzl J, Bein BK (1983) Z Phys Chem NF 134:17
56. Rosenczwaig A (Ed Marton L) In: Adv Electronics a Electron Physics, Vol 46. Academic Press, New York, pp 207–311
57. McChelland JF, Knisely RN (1976) Appl Opt 15:2658; Appl Phys Letters 28:467
58. Perkampus H-H (1982) Naturwissenschaften 69:162
59. Nordhaus O, Pelzl J (1981) Appl Phys 25:221
60. Rosenczwaig A (1980) Photoacoustics and Photoacoustic-Spectroscopy, Vol 57 d Serie Chem Anal. (Elving PJ, Wineforder JD (eds)) Wiley, New York
61. Yoh-han Pao (ed) (1977) Optoacoustic Spectroscopy and Detection. Academic Press, New York
62. Gliemeroth G, priv Mitteil
63. Katalog für Farb- und Filterglas, Schott Mainz, Nr 3525 (1974)
64. Görtz W, Perkampus H-H (1982) Fresenius Z Anal Chem 310:77
65. Helander P, Lundstroem J, McQueen D (1981) J Appl Phys 52:1146
66. Lin JW, Dudek LP (1979) Anal Chem 51:1627
67. Brücher H, Perkampus H-H (1985) Fresenius Z Analyt Chem 320:330
68. Tilgner R, Lüscher E (1978) Z Physik Chem NF 111:19
69. Meichenin D, Auzel F (1983) J Physique Suppl Fasc 10 C6–151
70. Roark JC, Palmer RA (1978) Chem Phys Lett 60:112
71. Plichon V, Cecile JL, Boissay S, Maillot M (1983) J Physique Suppl Fasc 10 C6–109
72. Breuer HD, Jacob H, Düster G (1982) Appl Opt 21:41
73. Breuer HD, Jacob H (1980) Chem Phys Lett 73:172
74. Breuer HD (1983) J Physique, Suppl Fasc 10 C6–321
75. Rosenczwaig A (1973) Opt Commun 7:305
76. Rosenczwaig A (1975) Anal Chem 47:5921; Phys Today 28:23 (1975)
77. Breuer HD (1980) Naturwissenschaften 67:91
78. Adams MJ, Beadle BC, King AA, Kirkbright GF (1976) Analyst 101:553
79. Junge K, Bein B, Pelzl J (1983) J Physique, Suppl Fasc 10 C6–55
80. Adams MJ, Kirkbright GF (1977) Analyst 102:670
81. Xiao Li, Brücher K-H, Görtz W, Perkampus H-H (1983) J Physique, Suppl Fasc 10 C6–137
82. Görtz W, Perkampus H-H (1982) Fresenius Z Analyt Chem 310:77
83. Helander P, Lundstroem J, McQueen D (1981) J Appl Phys 52:1146
84. McChelland JC, Knisely RN (1976) Appl Opt 13:2658
85. Teng YC, Royce BSH (1980) J Opt Soc Am 70:557
86. Görtz W (1982) Dissert Univers Düsseldorf
87. Schneider S, Coufal H, IBM Res Rep RJ 3352, 12/28/81
88. Schneider S, Coufal H (1982) J Chem Phys 76:2919
89. Schneider S, Möller U, Coufal H (1981) In: Coufal H, Korpiun P, Lüscher E (eds) Photoacoustic-Principles and Applications. Vieweg, Wiesbaden
90. Schneider S, Möller U, Alicka M (1983) J Physique, Fasc 10 C6–407
91. Poulet P, Chambron J, Unterreiner R (1980) J Appl Phys 51:1738
92. Wetsel GC (1982) J Photoacoust 1:33
93. Holter A, Perkampus H-H (1989) Ber Bunsenges Phys Chem 93:717
94. Holter A, Perkampus H-H (1989) Fresenius' Z Anal Chem 334:436

95. Görtz W, Perkampus H-H (1983) Z Naturforsch A 38a:1022
96. Görtz W, Perkampus H-H (1981) In: Coufal H, Korpiun P, Lüscher E (eds) Photoacoustic-Principles and Applications. Vieweg, Wiesbaden
97. Lahmann W, Ludewig HJ (1977) Chem Phys Lett 45:177
98. Adams MJ, Highfield JG, Kirkbright GF (1980) Anal Chem 52:1260
99. Adams MJ, Highfield JG, Kirkbright GF (1977) ibid 49:1850
100. Rockly MG, Wangh KM (1978) Chem Phys Lett 54:597
101. Cahen D, Garty H, Becker RS (1980) J Phys Chem 84:3384
102. Görtz W, Perkampus H-H (1983) Fresenius Z Anal Chem 316:180
103. Hess P (1983) In: Boschke FL (ed) Top Curr Chem Vol 111. Springer, Berlin Heidelberg New York
104. Murphy JC, Amondt LC (1977) J Appl Phys 48:3502
105. Bechthold PS, Campagna M, Chatzepetros J (1981) Optics Comm 36:369
Bechtold PS, Campagna M (1981) Optics Comm 36:373
106. Pelzl J, Klein K, Nordhaus O (1982) App Optics 21:94
107. Bechthold PS (1982) J Photoacoust 1:87
108. Breuer HD (1981) In: Coufal H, Korpiun P, Lüscher E (eds) Photoacoustic-Principles and Applications. Vieweg, Wiesbaden
109. Tam AC, Patel CKN (1979) Nature (London) 280:304
110. Hey E, Gollnick K (1982) J Photoacoust 1:1
111. Longworth JW (1971) Luminescence Spectroscopy. In: Lamola A (ed) Creation and Detection of the Excited State, Bd 1, Teil A, Kap 7. Dekker, New York, S 343ff
112. Perkampus H-H (1980) In: Kelker H (ed) Ullmanns Encyklopädie der techn Chem, 4. Aufl, Bd 5. Verlag Chemie, Weinheim, S 269ff
113. Eisenbrand J (1966) Fluorimetrie. Wiss Verlagsges, Stuttgart
114. Zander M (1981) Fluorimetrie. Anleit chem Laboratoriumspraxis, Bd 17. Springer, Berlin, Heidelberg New York
115. Kasha M (1950) Discuss Faraday Soc 9:14
116. Haller W, Perkampus H-H (1978) Ber Bunsenges Phys Chem 82:200;
Haller W (1977) Dissert Univ Düsseldorf
117. Perkampus H-H, Petermann G (1969) Ber Bunsenges Phys Chem 73:805
118. Kepler RG, Caris JC, Avakion P Abramson E (1963) Phys Rev Letters 10:400
119. Bettermann H (1983) Dissert, Univ Düsseldorf
120. Avatian P, Abramson E (1965) J Chem Phys 43:871
121. Dörr F (1966) Angew Chem 78:457
122. Dörr F (1971) Polarized Light in Spectroscopy and Photochemistry. In: Lamola AA (ed) Creation and Detection of the Excited State, Bd 1, Part A. Dekker, New York, S 53ff
123. Perkampus H-H, Knop JV, Knop A, Kassebeer G (1967) Z Naturforsch Teil A22, 1419
124. Parker CA (1968) Photoluminescence of Solutions with Applications to Photochemistry and Analytical Chemistry, Chap 5E. American Elseviers, New York, pp 438 ff
125. Lloyd JBF (1971) J Forensic Sci Soc 2:83, 153, 235
126. John Ph, Soutar I (1976) Anal Chem 48:520
127. Birks JB (1970) Photophysics of Aromatic Molecules. Wiley, New York
128. Warner IM, Christian GD, Davidson ER, Callis JB (1977) Anal Chem 49:564;
Warner IM, Callis JB, Davidson ER, Christian GD (1976) Clin Chem (Winston-Salem, NC) 22:1483
129. Wright JC, Tallant DR et al. (1979) Angew Chem 91:765
130. Perkampus H-H (1980) Ullmanns Encyklopädie der techn Chemie, 4 Aufl, Bd 5. Verlag Chemie, Weinheim
131. Michl J, Thulstrup EW (1986) Spectroscopy with Polarised Light. Verlag Chemie, Weinheim

6 Investigation of Equilibria

Anomalies in the Bouguer-Lambert-Beer law can generally be explained by equilibria, including association equilibria, and temporal changes of the system under measurement. Therefore, UV-VIS spectroscopy provides a method for following equilibria and the kinetics of chemical reactions. However, it is necessary that at least one component absorbs in the ultraviolet or visible spectral region.

6.1 General

If we formulate an equilibrium between substances X and Y to form XY, the thermodynamic equilibrium constant is:

$$K_a = \frac{a_{xy}}{a_x a_y} = \frac{c_{xy}}{c_x c_y} \cdot \frac{f_{xy}}{f_x f_y} = K_c \cdot \frac{f_{xy}}{f_x f_y} \; . \qquad (70/1)$$

If we assume that the extinction coefficients of substances X, Y and XY, available in an equilibrium, are large ($\geq 10^4\,l\,mol^{-1}\,cm^{-1}$) we can work with relatively low concentrations, i.e. an almost ideal solution. In this case, the activity coefficients would equal "unity" to a good approximation and an equilibrium constant can be determined unambiguously using the equilibrium concentration relationship. Thus, the determination of the equilibrium constant K_c originates in a static multicomponent analysis (see Sect. 4.2). The following equation applies generally to the measured absorbance A for a pathlength d (optical density D) and at the given initial concentrations c_{ox}, c_{oy}:

$$D = \frac{A_{\tilde{\nu}}}{d} = (c_{ox} - c_{xy})\,\varepsilon_x + (c_{oy} - c_{xy}) \cdot \varepsilon_y + \varepsilon_{xy} \cdot c_{xy} \; . \qquad (71/1)$$

Frequently, it is found that product XY shows an absorption spectrum which is shifted bathochromically with respect to the reactants X and Y. Therefore, a wavenumber $\tilde{\nu}$ or wavelength λ can be selected at which extinction coefficients ε_x and ε_y equal zero, i.e. the two product terms of Eq. (71/1) vanish. The concentration of product XY can then be easily determined if ε_{xy} is known; and an equilibrium constant can be established from the given initial concentrations of c_{ox} and c_{oy}.

The case is often encountered where one of the reactants X and Y does not absorb at all in the UV-VIS spectral region above 220 nm. Protolytic equilibria are such systems since the proton or H_3O^+ ion and the OH^- ion meet the above requirements.

6.2 Protolytic Equilibria; pK-Values

The following cases must be differentiated:

a) $BH^+ + H_2O \rightleftharpoons B + H_3O^+$,
b) $Ar-OH + H_2O \rightleftharpoons Ar-O^- + H_3O^+$,
c) $Ar-COOH + H_2O \rightleftharpoons ArCOO^- + H_3O^+$.

Case a) is the dissociation of a cationic acid which is formed with bases in acid solution.

Case b) is the dissociation of an acidic OH-group which we find for example in phenols.

Case c) is the dissociation of a carboxylic acid.

These cases are generally the ones where the absorption spectra of species BH^+, $Ar-O^-$ and $ArCOO^-$ are shifted bathochromically vis-a-vis the absorption spectra of species B, ArOH and ArCOOH. The protolytic equilibrium can be followed at a suitable position in the BH^+, ArO^- and $Ar-CO^-$ absorption spectra. Figure 49 shows an example of this for *p*-nitrophenol.

We use a general formulation of the dissociation equilibrium for describing cases a) to c):

$$BH + H_2O \rightleftharpoons B^- + H_3O^+ \ .$$

BH is the cationic acid (BH^+), phenol (ArOH) or carboxylic acid (ArCOOH).

For the concentrations in equilibrium the following always applies:

$$c_0 = c_{BH} + c_{B^-} \ . \tag{72/1}$$

If the degree of dissociation α is introduced, the ratio of concentrations c_{B^-}/c_{BH} at equilibrium is:

$$\frac{c_{B^-}}{c_{BH}} = \frac{\alpha}{1-\alpha} \ . \tag{72/1 a}$$

The absorbance A at specific wavelength λ is given by:

$$A_{exp} = (\varepsilon_{BH} \cdot c_{BH} + \varepsilon_{B^-} \cdot c_{B^-}) d \tag{73/1}$$

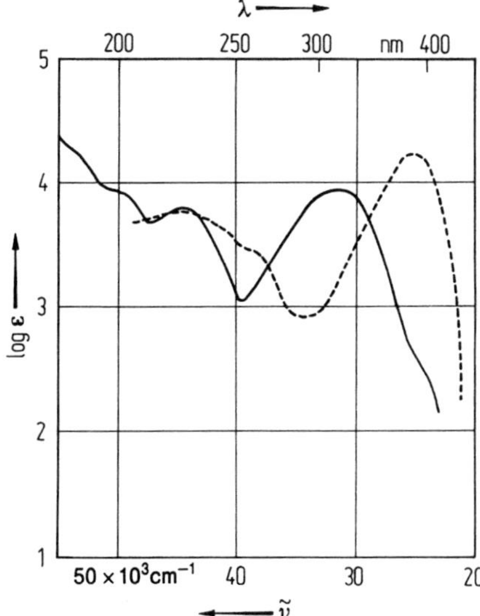

Fig. 49. Absorption spectrum of *p*-nitrophenol. H_2O (———), pH > 9 (- - -)

or

$$A_{exp} = A_{BH} + A_{B^-} \quad . \tag{73/1 a}$$

Concentrations c_{BH} and c_{B^-} of Eq. (73/1) can be expressed by means of Eq. (72/1), using the measured absorbance A_{exp} and the extinction coefficients of the pure species participating in the dissociation equilibrium (with d = 1 cm):

$$c_{B^-} = \frac{A_{exp} - \varepsilon_{BH} c_0}{\varepsilon_{B^-} - \varepsilon_{BH}}$$

and

$$c_{BH} = \frac{\varepsilon_B \cdot c_0 - A_{exp}}{\varepsilon_{B^-} - \varepsilon_{BH}} \quad . \tag{74/1}$$

The equilibrium constant of a protolytic reaction is defined as:

$$K_a = \frac{c_{B^-} f_{B^-}}{c_{BH} f_{BH}} \cdot \frac{a_{H_3O^+}}{a_{H_2O}} \quad . \tag{75/1}$$

After substituting the above expressions for c_B and c_{BH^+} in Eq. (75/1), and with $a_{H_2O} = 1$ and activity coefficients f_{B^-} and $f_{BH} = 1$ (dilute solution ca. 10^{-4} mol l^{-1}), we obtain:

$$K = \frac{A_{exp} - \varepsilon_{BH} c_0}{\varepsilon_{B^-} c_0 - A_{exp}} a_{H_3O^+} = \frac{\alpha}{1-\alpha} a_{H_3O^-} \tag{76}$$

or with

$$A_{exp}/c_0 = \varepsilon'_e$$

(The measured absorbance is always related to the total concentration c_0 at d – 1 cm)

$$K = \frac{\varepsilon'_e - \varepsilon_{BH}}{\varepsilon_{B^-} - \varepsilon'_e}\, a_{H_3O^+} = \frac{\varepsilon_{BH} - \varepsilon'_e}{\varepsilon'_{B^-} - \varepsilon_{B^-}}\, a_{H_3O^+} \quad . \tag{77}$$

For practical applications Eq. (75/1) is used in logarithmic form:

$$pK = \log \frac{\varepsilon_{BH} - \varepsilon'_e}{\varepsilon'_e - \varepsilon_{B^-}} + pH \tag{78}$$

or

$$\log \frac{\varepsilon_{BH} - \varepsilon'_e}{\varepsilon'_e - \varepsilon_{B^-}} = pH - pK \quad . \tag{78a}$$

Equation (78) or (78a) is the so-called Henderson-Hasselbach equation [1]. The dissociation constant defined in Eqs. (76, 77) is not a "thermodynamic dissociation constant" but a "mixed dissociation constant" since it connects those values which we can measure directly spectrophotometrically and electrochemically [2–4].

The connection between the mixed dissociation constant K and the classical K_c and the thermodynamic K_a is given by:

$$K = K_a \cdot \frac{f_{BH}}{f_{B^-}} = K_c \cdot f_{H_3O^+} = g(c_0) \quad . \tag{79}$$

By using the Debye-Hückel theory the thermodynamic dissociation constant K_a can be determined from Eq. (79) by extrapolation of c_0 to zero [5, 6].

The relationships developed here form the basis of the spectrophotometric titration method for determining the pK-values of *single-stage* dissociation systems.

The evaluation can be carried out using different methods. If ε_{BH} and ε_{B^-} in Eq. (78a) are known, the expression can be plotted as a function of the pH-value which can be set, for example, by means of buffer solutions.

A straight line is obtained which intersects the pH-axis at pH = pK. At this point:

$$\frac{\varepsilon_{BH} - \varepsilon'_e}{\varepsilon'_e - \varepsilon_{B^-}} = 1 \quad \text{and thus} \quad \varepsilon'_e = \frac{\varepsilon_{BH} + \varepsilon_{B^-}}{2} \quad . \tag{80}$$

Figure 50 shows an example of the *absorption spectra* of *p-nitrophenol* at different pH-values. The curve in Fig. 51 is plotted according to Eq. (78a). It is based on an evaluation of this protolytic equilibrium made by Blume and Lachmann [7]. The pK value found for *p*-nitrophenol is pK = 7.16±0.003. The type b) dissociation equilibrium with $\varepsilon_{BH} = \varepsilon_{ArOH}$ and $\varepsilon_{B^-} = \varepsilon_{ArO^-}$ applies to *p*-nitrophenol.

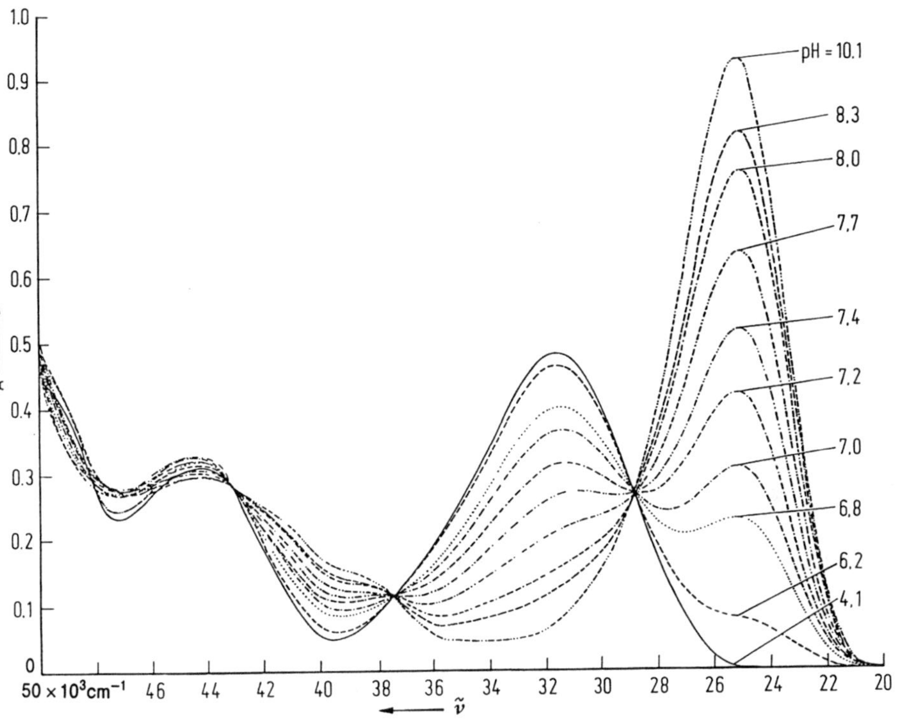

Fig. 50. pH-Dependence of the absorption spectrum of *p*-nitrophenol

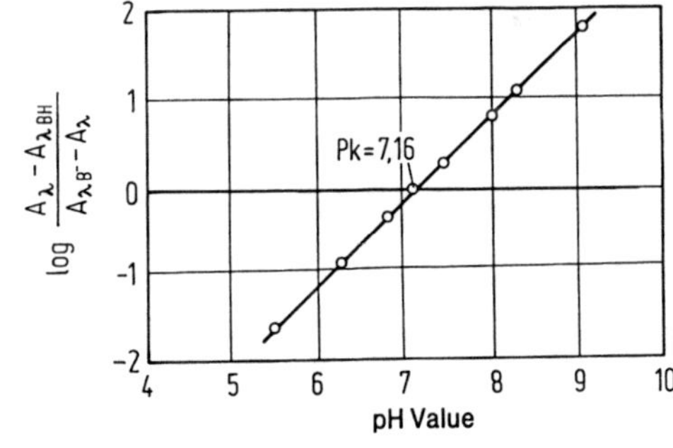

Fig. 51. Evaluation of the pH-dependence of the absorption spectrum of *p*-nitrophenol according to the Henderson-Hasselbach equation (78a)

For the determination of pK values by means of Eq. (78a), measurements must be taken at different pH-values. For that reason, a buffer solution must be prepared for every measurement and the sample under investigation (c_0) weighed in or a concentrated stock solution added. Although a widely used method, it has several disadvantages; it is very time-consuming

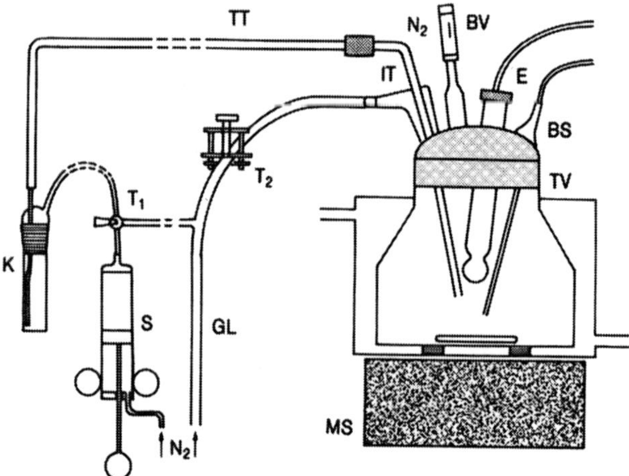

Fig. 52. Layout of a photometric titration apparatus designed by Lachmann and Polster (Sect. 6.1 [8])

and errors can occur when weighing or diluting a stock solution which can lead to variations of the total concentration c_0.

Lachmann and Polster have discussed this and numerous other *sources of error.* They have compared them with the advantages of a direct spectrometric titration [8]. Substance to establish the concentration c_0 is weighed *once* in this method. The required pH-value is set at each stage of the titration by the addition of base or acid. The sample is measured in a spectrometer and the next pH-value is then set.

Figure 52 shows an experimental arrangement for spectrometric titrations, where samples sensitive to oxygen can also be measured using N_2 as an *inert gas,* [8, 9]. The titration is carried out using burette BS in the actual titration vessel (TV) and the appropriate pH-value is measured with a glass electrode (E).

A modified Thunberg cuvette (K) with a special flow-through insert is used as the cuvette for absorption measurements. Teflon tubing (TT) connects the cuvette with inlet tube (IT). Cuvette, titration vessel and all tubes are flushed with pure nitrogen which flows through the gas line (GL) and taps (T1 and T2). As soon as an appropriate pH-value has been set by means of the burette and pH-meter the solution is slowly syphoned through the inert tubing (TT) and into the cuvette by means of the syringe (S).

The cuvette is thoroughly flushed by pumping the new solution in and out several times. The chamber behind the piston is also flushed with nitrogen in order to prevent external air penetrating the syringe at the rear. The whole titration vessel is mounted on a magnetic stirrer (MS) and can be temperature controlled.

A titration vessel with a large volume ($V_0 = 500$ ml) has the advantage that the dilution during a complete titration is only ca. 1 – 2 ml i.e., the error of dilution is <0.5%. Therefore, this error can be ignored in titration spectra recorded in analogue mode and sharp isosbestic points are maintained. For accurate quantitative evaluations, and in particular at small titration volumes (up to $V_0 \approx 20$ ml) the *dilution factor* must be eliminated mathematically [7, 8].

This mathematical correction is:

$$\frac{V_0+V_R}{V_0} = 1+\frac{V_R}{V_0}$$

where

V_0 = initial volume at the start of titration,
V_R = total volume of titrant up to the step in question.

In addition to the evaluation according to Eq. (78a) which has already been discussed, several other methods are known for determining pK-values. In a single-stage dissociation reaction, the measured absorbance $A_{\tilde{\nu}}$ as a function of the pH-value shows a sigmoid (S-shaped) curve.

Solving Eq. (77) for ε'_e we obtain:

$$\varepsilon'_e = \frac{\varepsilon_{B^-} K + \varepsilon_{BH} a_{H_3O^+}}{a_{H_3O^+} + K} = \frac{\varepsilon_{B^-} 10^{-pK} + \varepsilon_{BH} 10^{-pH}}{10^{-pH} + 10^{-pK}} \tag{81}$$

or with

$$\varepsilon'_e = A_g/c_0;\ \varepsilon_{B^-} = A_{B^-}/c_0 \quad \text{and} \quad \varepsilon_{BH} = A_{BH}/c_0 (d = 1\ \text{cm!})$$

$$A_e = \frac{A_{B^-} 10^{-pK} + A_{BH} 10^{-pH}}{10^{-pH} + 10^{-pK}} \ . \tag{81a}$$

S-shaped spectrometric titration curves can be calculated from Eq. (81a) and the determination of the *point of inflexion* in order to obtain the pK-value is based on the application of this equation. However, an inflexion point analysis requires that the points of the titration curve are symmetrical with respect to those of the pK-value to be determined, i.e. the pK-value and inflexion point are identical [10]. The tangent method [10], ring method [11] and differential technique [6, 10] can be used for determining the inflexion point accurately. When using these methods information about the extreme values ε_{B^-} and ε_{BH} of the titration is not necessary.

If these two values are known, the mid-point of the total extinction change can be taken as shown in Eq. (80) since the value associated with ε'_e corresponds to the pK-value on the pH-scale. Other methods utilize a linearization of the titration curves.

We have discussed one of these methods in connection with the Henderson-Hasselbach equation, (78a). Equation (77) forms the basis for another technique, from which we obtain the expression:

$$(\varepsilon'_e - \varepsilon_{BH}) \cdot a_{H_3O^+} = K \cdot (\varepsilon_{B^-} - \varepsilon'_e) \quad \text{or}$$

$$(\varepsilon'_e - \varepsilon_{BH}) \cdot 10^{-pH} = -K\varepsilon'_e + K\varepsilon_{B^-} \ . \tag{82}$$

If the left-hand side is plotted against ε_e' or A_{exp} a straight line is obtained whose slope yields K directly.

The determination of pK-values of single-stage dissociation systems can be modified. Using Eq. (75/1) as a basis we can write:

$$pK = pH + \log \frac{c_{BH}}{c_{B^-}} + \log \frac{f_{BH^+}}{f_B} . \tag{83}$$

If it is also assumed initially that the activity coefficients $f_{BH} = f_{B^-} = 1$, then $\log (c_{BH}/c_{B^-}) = 0$ at pK = pH, i.e. $c_{BH} = c_{B^-}$. Thus, if an equimolar solution of BH and B^- is prepared it should have a pH-value which corresponds to the pK-value. In many cases, this requirement can only be met approximately. However, if the concentration ratio c_{BH}/c_{B^-} is measured spectrophotometrically in such a solution and simultaneously the pH-value of the solution is established we can very easily determine accurate pK-values since the c_{BH}/c_{B^-} ratio can be determined most accurately spectrophotometrically at pH ≈ pK. However, the extinction coefficients ε_{BH} and ε_{B^-} must be known when using this method, see Eq. (74/1). Basically, this method can be related to the inflexion point analysis and to spectrometric titration in the proximity of the inflexion point. Perkampus and Prescher have determined the *pK-values of pyridine and its methyl derivatives* in this manner. They took account of the influence of the activity coefficient f_{BH^+} in Eq. (83) using the Debye-Hückel equation [12]. Since the effect of temperature can also be measured easily in this way, all the data for describing protolytic equilibria are accessible using known thermodynamic relationships.

All the above methods have the disadvantage that the reliability of the pK-values depends on the conventional pH-scale. However, for a single-stage dissociation reaction it can be shown that a *purely optical determination* of the pK-value is possible.

From the three types of dissociation mentioned above, the following applies to a phenol dissolved in water, taking into account the requirement of electrical neutrality:

$$c_{H_3O^+} = c_{ArO^-} = \alpha \cdot c_0 ,$$

$$K = \frac{c_{H_3O^+} \cdot c_{ArO^-}}{c_{ArOH}} = \frac{c_{ArO^-}^2}{c_0 - c_{ArO^-}} = \frac{\alpha^2}{1-\alpha} c_0 \tag{84}$$

where α is the degree of dissociation of the phenol or other species. When using this relationship it is also assumed that there exists a wavelength region in which the pure acid such as the phenol does not absorb, i.e. $\varepsilon_{ArOH} = 0$.

It can be seen in Fig. 49 that this requirement is met for *p-nitrophenol.* At a given concentration c_0 and pathlength d and with knowledge of the extreme value A_{ArO^-} in alkali, the concentration c_{ArO^-} is easily accessible from the absorption measurements on a pure aqueous phenol solution. Therefore, α and K can be calculated via Eq. (84). If the known concentration c_0 is varied in several experiments the thermodynamic dissociation constant K_a can be determined by extrapolating to ionic strength $I = 0$ by means of the Debye-Hückel equation [5, 6].

This technique is not a true titration method since only two titration points are checked photometrically, i.e.

1. the fully alkaline solution
2. the pure aqueous solution.

Thus, a precise measurement of the concentration c_0 and of the extinction coefficient ε_{ArO^-} is carried out instead of a pH-measurement; and the problems of a conventional pH-scale are avoided.

If the requirement of $\varepsilon_{ArOH} = 0$ is not met in the wavelength region under investigation the degree of dissociation α can be determined by an approximate method if ε_{ArOH} is known [13]. Kortüm and Shih have carried out detailed and thorough investigations of the photometric determination of dissociation constants of single-stage dissociation equilibria in mixtures of solvents [14].

The foregoing considerations demonstrate that, on account of the limits imposed by stoichiometry, spectrometric titration systems are always described by means of a single *concentration variable*, i.e. c_{BH} or c_{B^-}.

From Eq. (73/1) (with $c_{B^-} = c_0 - c_{BH}$), the following equation applies to the absorbance measured at wavelength '1':

$$\begin{aligned} A_1 &= \varepsilon_{1.BH} c_{BH} d + \varepsilon_{1.B^-} c_{B^-} d \\ &= \varepsilon_{1.BH} c_{BH} d + \varepsilon_{1.B^-} c_0 d - \varepsilon_{1.B^-} c_{BH} d \ . \end{aligned}$$

If the constant expression $\varepsilon_{1.B^-} c_0 d = A_{1.B^-}$ is taken over to the left-hand side we obtain:

$$\Delta A_1 = A_1 - A_{1.B^-} = (\varepsilon_{1.BH} - \varepsilon_{1.B^-}) c_{BH} d \tag{85}$$

and analogously we obtain for wavelength "i":

$$\Delta A_i - A_{i.B^-} = (\varepsilon_{i.BH} - \varepsilon_{i.B^-}) c_{BH} d \ . \tag{85a}$$

If Eq. (85) is divided by Eq. (85a) the equation for the *absorbance-difference (ΔA) diagram* results:

$$\frac{\Delta A_1}{\Delta A_i} = \frac{\varepsilon_{1.BH} - \varepsilon_{1.B^-}}{\varepsilon_{i.BH} - \varepsilon_{i.B^-}} = Z_i \ , \tag{86}$$

$$\Delta A_1 = Z_i \Delta A_i \ . \tag{86a}$$

Z_i is a constant formed from the extinction coefficients of the cationic acid (or phenol) and the base (or phenolate ion) at wavelengths λ_i and λ_1.

According to this equation, a ΔA-diagram should consist of straight lines passing through the origin with slopes given by Z_i. Figure 53 shows the ΔA-diagram for p-nitrophenol at eight wavelength combinations [7, 8]. We see a good agreement of the linearity up to $pH \leq 9$. A deviation occurs in

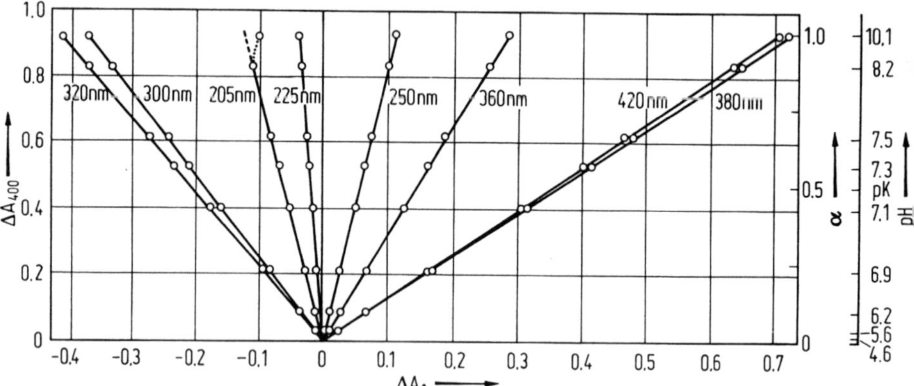

Fig. 53. Absorbance-difference diagram (ΔAD) for Fig. 50; from Lachmann [8]

the line for $\lambda_i = 205$ nm which can be ascribed to the hydroxyl ion, the absorption of which becomes apparent at this wavelength. Therefore, the ΔA-diagram can be used to check whether we are dealing with a uniform or single-stage dissociation equilibrium. The diagram taken in conjunction with the absorption spectra also facilitates selection of the most favorable photometric measuring range.

In addition to a ΔA-diagram, measurements can also be represented by means of an *absorbance diagram.* If A_{1,B^-} is added to both sides of Eq. (86a) then since $\Delta A_1 = A_1 - A_{1.B^-}$ by Eq. (85), there results

$$A_1 = Z_i A_i - Z_i A_{i.B^-} + A_{1.B^-} \ ,$$

$$A_1 = Z_i A_i + c_0 d \,|\, \varepsilon_{1.B^-} - Z_i \cdot \varepsilon_{i.B^-} \,| \ . \tag{87}$$

Plotting A_1 as a function of A_i also results in a straight line which, in contrast to the ΔA-diagram, does not pass through zero [9]. It should be noted that parameters ΔA_i and A_i are of course related as parameters to the variable pH-values. For that reason, the pH-value has been plotted as a non-linear auxiliary scale in the right-hand margin together with the degree of dissociation α which results directly from $c_{BH}/c_{B^-} = \alpha/1-\alpha$. See Eq. (72/1a).

Equation (86) shows directly the existence of *isosbestic points.* If $\varepsilon_{1.BH} = \varepsilon_{1.B^-}$ applies then there is no change of absorbance throughout the whole titration. Therefore, if the titration spectra intersect they must always form sharp isosbestic points.

Equations (86, 86a) and (87) represent the criteria for graphical matrix rank analysis for factor s = 1 [8, 9]; see Sect. 7.2.

We frequently have to deal with two- and multi-step titration systems in the *spectrophotometric investigation of dissociation equilibria.* If the individual dissociation equilibria overlap slightly in such systems they can be divided into single-stage subsystems. As a guide, it can be assumed that for

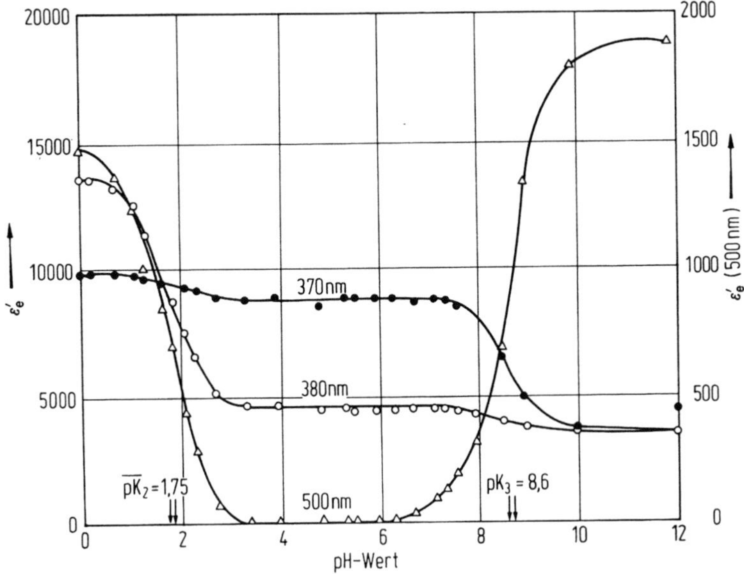

Fig. 54. Extinction as a function of the pH-value for three different wavelengths; 4-hydroxy-phenazine

pK-value differences of up to 3.5 to 4 units overlapping cannot be detected because current measuring techniques are not sufficiently accurate. In this case, the previously discussed methods can be used for the evaluation of the individual subsystems. A-diagrams form a good criterion for absence of overlap. By Eq. (87) they yield intersecting straight lines for each subsystem [15]. Lachmann has discussed these relationships in detail using 3-desoxy-pyridoxol as the example [9].

Hydroxyderivatives of phenazine are in principle three-step dissociation systems; and the following equilibria apply to the stages:

$$\text{Dication} \xrightleftharpoons{K_1} \text{monocation} \xrightleftharpoons{K_2} \text{base} \xrightleftharpoons{K_3} \text{anion} \;.$$

However, only stages K_2 and K_3 are observed in the pH-region$\geqslant 0$ because K_1 lies at higher acidities in the region covered by the Hammett function [16, 17]. Figure 54 shows the A(pH)-curves at three wavelengths. The considerable pK interval between stages K_2 and K_3 is clearly recognizable. This permits an evaluation without difficulty using the inflexion point method or Eq. (78a) [16] and we obtain:

$$pK_2 = 1.75 \;, \qquad \Delta pK = 6.85 \;. \qquad pK_3 = 8.60 \;.$$

Titration spectra of multi-stage dissociation systems are difficult to evaluate if their pK-values lie close together, i.e. if ΔpK is less than 3. Numerous techniques for analyzing such systems have been developed in recent years. Polster [18–21] and Lachmann [8, 9, 22] have shown that a systematic use and evaluation of the absorbance diagrams, absorbance-difference diagrams and absorbance-difference-quotient diagrams are of particular value. A graphical evaluation of all subsections of a multi-stage spectrophotometric titration is very time-consuming. For that reason, a nonlinear curve-fitting technique which takes account of *all* A(pH)-curves has been developed for routine evaluation of multi-step titration systems [23]. Such methods for the evaluation of individual A(pH)-curves were previously known [24–26].

Since the extinction coefficients of all species cannot in general be determined experimentally, it is usually assumed that the pK-values and all coefficients are unknown factors [8, 9]. Absorbance values, $A_\lambda(pH)$, measured at n wavelengths and j pH-values are known which in principle can be represented in terms of the initially unknown extinction coefficients, $\varepsilon_\lambda(pH)$, and pK-values. Thus, $A_{\lambda,exp}$ and $A_{\lambda,calc}$ can be compared with each other at each iteration stage or the minimum of the error function can be found by using the least squares method.

Of course, this nonlinear curve-fitting technique is also suitable for optimizing single- and multi-step *non-overlapping* titration systems. However, its superiority is most apparent for multi-step overlapping titration systems as shown by Lachmann [9].

The following pK-values of phthalic acid (two-steps) have been found by this method:

$$pK_1 = 2.92 \pm 0.01 \ , \qquad \Delta pK = 2.42 \ .$$
$$pK_2 = 5.34 \pm 0.01 \ ,$$

Benzene-1,2,4-tricarboxylic acid yields the following values:

$$pK_1 = 2.53 \pm 0.01 \ , \qquad \Delta pK = 1.44 \ .$$
$$pK_2 = 3.97 \pm 0.01 \ , \qquad \Delta pK = 1.44 \ .$$
$$pK_3 = 5.41 \pm 0.01 \ ,$$

6.3 Complex-Forming Equilibria

A new chemical substance is formed in a chemical equilibrium. Generally, we can isolate the new compound, which is held together by the principal valence forces, and accurately determine its physical and chemical properties. If the compound incorporates a chromophoric system then the UV-VIS absorption spectrum can also be assumed to be known.

In many cases, this assumption cannot be made for complex-forming or association equilibria. Therefore, the quantitative investigation of the UV-

VIS spectrum is frequently difficult. Even if complexes or molecular compounds can be isolated in a *solid* state they tend to dissociate in solution on account of the weak bonding forces.

Therefore, the extinction coefficient of a stoichiometrically defined complex (molecular compound, ion associate) is not usually directly accessible in solution.

Generally, complex formation in solution can be described by the following reaction equation:

$$mD + n \cdot A \rightleftharpoons D_m A_n \ . \tag{88}$$

Since every type of interaction leading to the formation of a stoichiometrically composed complex or associate can be characterized as a donor-acceptor interaction, we write D for a donor and A for an acceptor in Eq. (88).

Depending on the specific type of this interaction, we can differentiate between proton donors and acceptors as well as electron donors and acceptors. In the former, association equilibria of a hydrogen bond are involved and in the latter formation equilibria of molecular compounds, which are also called electron donor-acceptor (EDA) complexes or charge-transfer (CT) complexes. Numerous complexes between metal cations and inorganic or organic ligands can be assigned to EDA-interactions.

In addition to the equilibrium (88), we must also consider self-association:

$$n \cdot M \rightleftharpoons M_n \ . \tag{89}$$

$n = 2$ is a dimerisation. If n increases considerably this is a polymeric association, as discussed by Scheibe for the cations of polymethine dyes [27, 28].

A detailed presentation of every type of complex formation or association phenomenon cannot be given here.

6.3.1 H-Bond Association

When considering the association equilibria of an H-bond, we distinguish association between like molecules (Eq. 89) or between different molecules (Eq. 88).

$n = m = 1$ is the simplest case of an H-bond association of different molecules. In the association of like molecules, apart from $n = 2$, there are also multiple associations $n \geq 3$ dependent on concentration. Alcohols have been preferred for the investigation of an association of like molecules. Since the effect on the OH-valence vibration is the characteristic parameter here, such investigations have been carried out almost exclusively in the IR spectral region. Since the pioneering work of Mecke and Kempter [29], an indeterminable number of papers concerning hydrogen bonding have been published [30].

The *association of different molecules* has also been frequently investigated in the IR spectral region. The effect on the OH-valence vibration of the proton donor is used as the indicator. However, it must be ensured that the acceptor does *not* absorb in the region of an OH-valence vibration. Furthermore, the concentration of the donor must be kept so small that an association of like molecules can be ignored or can be separately determined [31–33].

In contrast to IR and NMR investigations, measurements in the UV-VIS spectral region and their *quantitative* evaluation are relatively rare. The reasons for this may be that a characteristic feature such as an OH-valence vibration, available in vibrational spectroscopy, does not occur in electronic excitation spectra. However, if unsaturated N-heterocycles are used as proton acceptors the characteristic shifts of their absorption spectra in the UV-VIS spectral region can be assigned to hydrogen bonding.

The best known among these are proton-donor-acceptor interactions where the acceptor has an $n \rightarrow \pi^*$ transition that is shifted hypsochromically by the interaction [34–36].

These phenomena were first observed a long time ago [37] and are not limited to aza-aromatic compounds. They are apparent when the spectra of a specific substance in aprotic solvents are compared with those of the same substance in protic solvents. Such shifts of bands also occur in the UV-VIS spectrum if the proton donor is added to the solution of an aza-aromatic

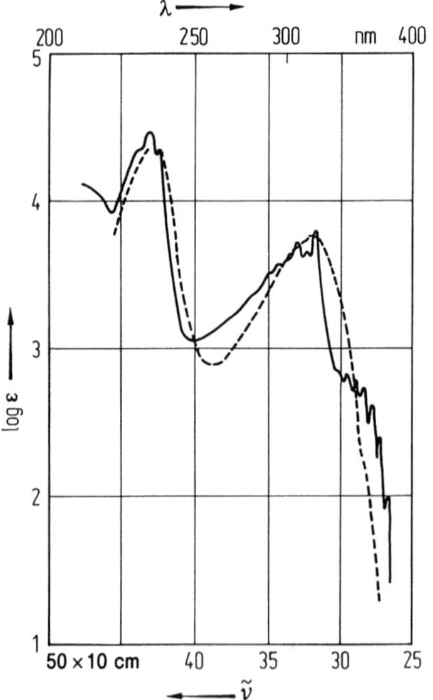

Fig. 55. Absorption spectrum of quinoxaline. *n*-heptane (———), H_2O (– – –)

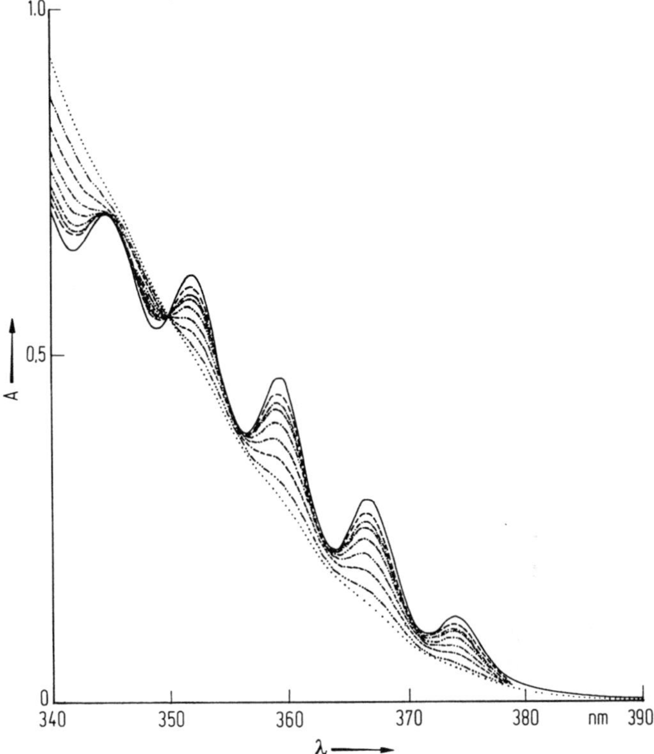

Fig. 56. $n-\pi^*$ band of quinoxaline in *iso*-propanol/*n*-heptane mixtures as a function of the concentration of *iso*-propanol (*full line*, *n*-heptane; *dotted lines*, increasing concentration of *iso*-propanol)

substance in an inert solvent rather than being used directly as the solvent [38, 39].

Figure 55 shows the change of the UV spectrum of *quinoxaline* on changing the solvent from *n*-heptane to water (pH 6, phosphate buffer) [40]. The H-bond association with quinoxaline has been followed spectrophotometrically as a function of the alcohol concentration in the solvent system *iso*-propanol/*n*-heptane [41].

In addition to quinoxaline, substituted quinoxalines were used in order to trace the influence of methyl substitution on the acceptor strength. Characteristic isosbestic points were always observed which indicated the presence of association equilibria (Fig. 56).

The spectra illustrated in Fig. 56 as a function of the alcohol concentration correspond formally to titration spectra discussed in the previous section. By analogy with the Henderson-Hasselbach equation (78, 78a), we obtain for such an association equilibrium:

$$B + nA = BA_n \ , \tag{90}$$

$$\log K = \log \frac{c_{ass}}{c_B} - n \log c_A = \log \frac{\varepsilon_B - \varepsilon'_e}{\varepsilon'_e - \varepsilon_{ass}} - n \log c_A \tag{91}$$

where

ε_B is the extinction coefficient of unassociated base molecules,
ε_{ass} is the extinction coefficient of the pure associates (BA_n),
ε'_e is A_{exp}/c_0 with c_0 = total concentration of the base (d = 1 cm),
c_A is the alcohol concentration; c_B and c_{ass} are the concentration of base and associate in equilibrium.

In the quinoxaline/*iso*-propanol system, a straight line with a slope of $n = 1$ results when plotting $\log(\varepsilon_B - \varepsilon'_e)/(\varepsilon'_e - \varepsilon_{ass})$ against $\log c_A$ for alcohol concentrations c_A not above 1 mol l^{-1}. This means that a 1:1 H-bonded complex is formed at low alcohol concentrations.

Brealey and Kasha [42] have carried out parallel investigations using pyridazine and ethanol. An evaluation using Eq. (91) again provides a straight line with a slope of $n = 1$ at low ethanol concentrations; $c_A < 1 \text{ mol l}^{-1}$. Thus, a 1:1 complex is again formed.

The $n \rightarrow \pi^*$ band of pyridazine is shifted hypsochromically by 1450 cm^{-1} in this concentration range. An additional hypsochromic shift of a further 1000 cm^{-1} occurs above this alcohol concentration up to pure alcohol, which is comparatively small vis-a-vis a shift of 1450 cm^{-1}.

Since an isosbestic point is not observed in this set of spectra, other competing interactions must be considered, as well as the possible formation of a second H-bond (Lippert [43]).

In contrast to the quantitative examples mentioned, numerous investigations of hydrogen bonding have been carried out in the UV-VIS region to classify bands. $n \rightarrow \pi^*$ bands are displaced towards the blue and $\pi \rightarrow \pi^*$ bands are displaced towards the red compared with their position in n-hexane and similar aprotic solvents. Pimentel has shown that the band shifts, $\Delta\tilde{\nu}_a$, in cm^{-1} correlate with the H-bond energies i.e. their differences in the ground and excited state as well as with the Franck-Condon excitation energy [44].

Such investigations, where an excess of the proton donor is added or where it is used directly as a solvent, can be easily and simply carried out. Generally, they also show the essential changes detectable by spectroscopy; although only a final value is involved e.g. a displacement towards the red, $\Delta\tilde{\nu}_a$, of a $\pi \rightarrow \pi^*$ band. Such measurements are not suitable for a more accurate evaluation of intensities or vibrational fine structure since the molecular structure of the complexes giving rise to the spectra cannot be defined precisely.

At a lower proton-donor concentration the acceptor is not completely complexed and the spectrum of the free base overlies that of the H-bonded complex of the acceptor.

Although a larger proton-donor excess leads to more complete complexing of the acceptor, it also gives rise simultaneously to further association which can have a considerable effect upon the spectrum.

Therefore, it is appropriate to use the spectrum of a single association structure, as accurately defined as possible, for a comparison with the spectrum of the free acceptor. It is best to use the simplest structure DH ... A which can be realized in a dilute solution in an inert solvent with monoazaaromatic substances as acceptors [45].

Among the bases investigated as acceptors are pyridine, quinoline, isoquinoline, 5,6-benzquinoline, 5,6-benzoisoquinoline, acridine, 2,3-benzacridine and 1,2,7,8-dibenzacridine.

With 1,1,1,3,3,3-hexafluoro-isopropanol (HFIP) as proton donor, the extinction measured in any mixture is given by:

$$\frac{A_{exp}(\lambda)}{d} = \varepsilon_B(\lambda)\cdot c_B + \varepsilon_K(\lambda)c_K ; \tag{92}$$

where

ε_B is the extinction coefficient of the pure base,
ε_K is the extinction coefficient of the pure 1:1 complex,
c_B, c_K are the equilibrium concentrations of base and complex.

As illustrated in Sect. 5.2 with the example of isoquinoline, a calibration line for the base concentration can be drawn using first order derivative spectroscopy. Therefore, the equilibrium concentration c_B of the base can be determined for each HFIP concentration. This also provides the concentration c_K from $c_0 = c_B + c_K$ and yields the extinction coefficients of the H-bonded complexes:

$$\varepsilon_K(\lambda) = \frac{1}{c_K}\left[\frac{A_{exp}(\lambda)}{d} - \varepsilon_B(\lambda)c_B\right] . \tag{93}$$

The presence of *one* (spectroscopically) uniform association type is confirmed by the fact that completely identical spectra $\varepsilon_K(\lambda)$ for a hydrogen-bonded complex are obtained at different degrees of complexation (up to 80%) [45]. Figure 57 shows the long-wavelength band of the H-bonded complex of 1,2,7,8-dibenzacridine-HFIP in *n*-heptane in comparison with the free base. As is to be expected in these systems, the spectrum of the complex will be recognized as shifted bathochromically since a $\pi \rightarrow \pi^*$ transition is being observed.

The complex spectra of the other bases mentioned were determined analogously (for a discussion see [45, 46]).

Since the concentration of the complex, c_K, at the given gross concentration c_0 is obtained via the equilibrium concentration, c_B, the association constant $K_{c,T}$ can be determined [47]. Due to the relatively high extinction coefficients of N-heterocycles, UV-VIS measurements are possible at relatively low concentrations of the bases ($c_{0B} > 10^{-5}$ mol l^{-1}). Therefore, to a good approximation, we are dealing with an ideal solution.

The H-bonding association of *1,2,7,8-dibenzacridine* in carbon tetrachloride has been investigated fully [47]. CCl_4 was selected in order to

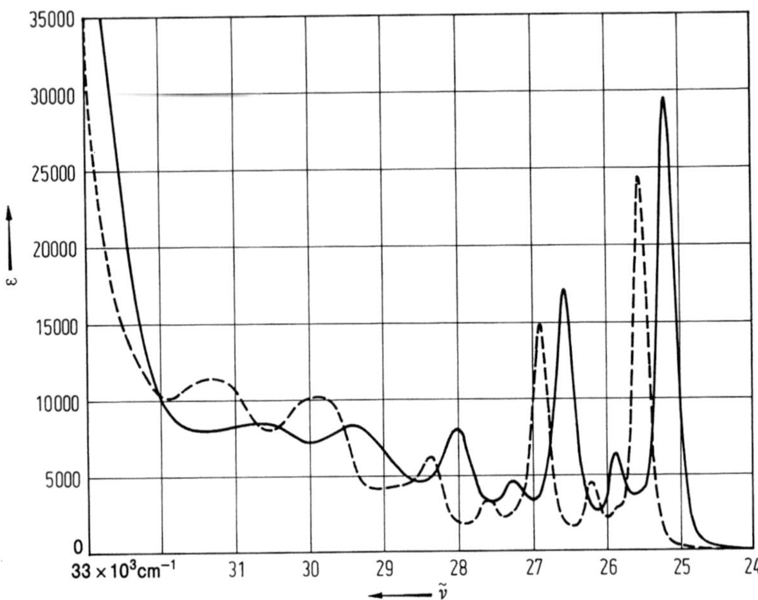

Fig. 57. Long-wavelength absorption bands of 1,2,7,8-dibenzacridine in *n*-heptane (– – –); H-bonded complex with hexafluoroisopropanol 1/1 (———) in *n*-heptane

Table 15. Association constants for 1,2,7,8-dibenzacridine (DBA) with proton donors in CCl_4; 293 K; comparison of IR and UV-VIS spectra data [47]

Proton donor	K_c $\|l\,mol^{-1}\|$	
	IR	UV-VIS
Phenol	51.0	49.7
4-fluorophenol	79.8	84.4
4-chlorophenol	123.0	124.0
4-bromophenol	132.0	132.0
3,4,5-trichlorophenol	578.0	569.0
1,1,1,3,3,3-hexafluoroisopropanol	257.0	277.0
2,2,2-trifluoroethanol	~25.0*	23.8

* When studying the system trifluoroethanol-DBA in the IR region, we require extremely high base concentrations (up to saturation). Thus, an association constant cannot be determined accurately from IR measurements.

allow a comparison of the results with investigations in the IR spectral region. Table 15 shows, within the accuracy of measurement, an excellent agreement of the K_c values.

Since the K_c values are a function of temperature, the relevant thermodynamic data for the association equilibria can be determined [47].

The excellent agreement between IR and UV-VIS measurements illustrates a possibility not available with IR investigations of H-bonding association. CCl_4, and occasionally tetrachloroethylene, have been used almost exclusively as solvents in IR measurements but in the UV-VIS region, different solvents can be used for the measurements. *n*-Hexane or *n*-heptane are ideal solvents since they are "standard solvents" which cause the least intermolecular interactions and exert the minimum influence on association equilibria.

The K_c-values in CCl_4, *n*-heptane, benzene and toluene show the extremely strong effect of the solvent. This is not only noticeable in the absolute values of the association constants, but also in the bond enthalpies of the hydrogen bond. The association constants are considerably larger in *n*-heptane than in CCl_4 or benzene; the latter is an extreme case where very small K_c-values are found. We notice here the specific interactions of the solvents with the dissolved proton donors which, as competing processes, partake in the equilibria [48, 49].

The examples show that UV-VIS spectroscopy can be used for quantitative investigation of H-bond association equilibria. For a mixed association with those proton acceptors which show intense $\pi \rightarrow \pi^*$ transitions it is possible to work with extremely low concentrations in comparison with those used in IR spectroscopy. When alcohols are used as proton donors, there is again a favorable situation because one component of the association equilibrium does not absorb in the UV-VIS spectral region. Finally, there is also the opportunity to work with *n*-heptane or *n*-hexane. This provides the great advantage of reducing to a minimum the specific interactions between the solvent and the dissolved substance.

6.3.2 EDA Complexes

Electron donor-acceptor (EDA) or *charge-transfer* (CT) complexes occupy a special position among molecular complexes. They frequently form as stoichiometrically defined 1:1 complexes in solution and are characterized by broad and structureless absorption bands having extinction coefficients between 500 and 20000 [$l\,mol^{-1}\,cm^{-1}$] shifted to long wavelengths from the absorption of the donor and acceptor components themselves [50].

This fact permits spectrophotometric investigation of *formation equilibria* and the determination of the equilibrium constants and extinction coefficients of the complexes.

Using equilibrium (88), we obtain for K_c

$$K_c = \frac{c_{D_mA_n}}{c_D^m \cdot c_A^n} = \frac{c_{D_mA_n}}{(c_{0D} - m\,c_{D_mA_n})^m (c_{0A} - n\,c_{D_mA_n})^n} \,. \tag{94}$$

c_{0D} and c_{0A} are the initial concentrations of donor and acceptor; $c_{D_mA_n}$ is the equilibrium concentration of the complex.

A 1 : 1 complex is the most common complex and we can simplify equation (94) with m = n = 1:

$$K_c = \frac{c_{DA}}{(c_{0D}-c_{DA})(c_{0A}-c_{DA})} \quad . \tag{94a}$$

The measured absorbance A_{exp} (= A_e) at a selected wavelength is due to the equilibrium composition:

$$A_e = \varepsilon_D \cdot (c_{0D}-c_{DA}) \cdot d + \varepsilon_A (c_{0A}-c_{DA}) \cdot d + \varepsilon_{DA} \cdot c_{DA} \cdot d \quad . \tag{95}$$

From this we obtain for c_{DA}:

$$c_{DA} = \frac{A_e - (\varepsilon_D c_{0D} + \varepsilon_A c_{0A}) d}{(\varepsilon_{DA} - \varepsilon_A - \varepsilon_D) \cdot d} = \frac{\Delta A}{\Delta\varepsilon \cdot d} = \frac{D'}{\Delta\varepsilon} \tag{96}$$

with

$$\Delta A = A_e - \varepsilon_D c_{0D} d - \varepsilon_A \cdot c_{0A} d = A_e - A_D - A_A$$

or

$\Delta A/d = D'$ as optical density

and

$$\Delta\varepsilon = \varepsilon_{DA} - \varepsilon_A - \varepsilon_D \quad .$$

In order to obtain a linear relation for an evaluation, we invert Eq. (94a)

$$\frac{1}{K_c} = \frac{c_{0D} \cdot c_{0A}}{c_{DA}} - (c_{0D} + c_{0A}) + c_{DA} \quad . \tag{97}$$

Since we usually work with a donor excess in these investigations, $c_{0D} + c_{0A} \gg c_{DA}$ applies to a good approximation.

If we take account of this and write c_{DA} as shown in Eq. (96) we obtain the relationship:

$$\frac{1}{K_c} = \frac{c_{0D} c_{0A} d \Delta\varepsilon}{\Delta A} - (c_{0D} + c_{0A}) = \frac{c_{0D} c_{0A} \Delta\varepsilon}{D'} - (c_{0D} + c_{0A})$$

and after transformation and conversion:

$$\frac{c_{0D} c_{0A} d}{(c_{0D} + c_{0A}) \Delta A} = \frac{c_{0D} c_{0A}}{c_{0A} + c_{0D}) D'} = \frac{1}{K_c \Delta\varepsilon} \cdot \frac{1}{c_{0D} + c_{0A}} + \frac{1}{\Delta\varepsilon} \quad . \tag{98}$$

If the left-hand side is plotted against $1/c_{0D} + c_{0A}$, $(K_c \Delta\varepsilon)^{-1}$ is obtained from the slope of the straight line and $(\Delta\varepsilon)^{-1}$ from the ordinate intercept.

If the extinction coefficients of donor (ε_D) and acceptor (ε_A) are known, ε_{DA} and K_c can be obtained.

As noted above, due to the low solubility of acceptors a donor excess is used. Therefore, it is often the case that $c_{0D} \gg c_{0A}$. Equation (98) then becomes the Ketelaar equation [59] Eq. (99):

$$\frac{c_{0A}\cdot d}{\Delta A}=\frac{c_{0A}}{D'}=\frac{1}{K_c c_{0D}\cdot\Delta\varepsilon}+\frac{1}{\Delta\varepsilon}\;. \tag{99}$$

If it is assumed that the donor does not absorb in the spectral region under investigation ($\varepsilon_D = 0$), then $\Delta A = A_e - A_A$ and $\Delta\varepsilon = \varepsilon_{DA} - \varepsilon_A$; thus Eq. (99) can be written as follows:

$$\frac{1}{\varepsilon'-\varepsilon_A}=\frac{1}{K_c c_{0D}(\varepsilon_{DA}-\varepsilon_A)}+\frac{1}{\varepsilon_{DA}-\varepsilon_A} \tag{99a}$$

with

$$\varepsilon'=\frac{A_e}{c_{0A}d}\quad\text{and}\quad\varepsilon_A=\frac{A_A}{c_{0A}d}\;.$$

This may be plotted and evaluated by analogy with Eq. (98).

It is frequently the case that donor *and* acceptor do not absorb in the spectral region in question ($\varepsilon_A = \varepsilon_D = 0$). Under these conditions Eq. (99) is transformed into the well-known Benesi-Hildebrand equation [52] Eq. (100):

$$\frac{1}{\varepsilon'}=\frac{c_{0A}d}{A_{exp}}=\frac{c_{0A}}{D}=\frac{1}{K_c c_{0D}\cdot\varepsilon_{DA}}+\frac{1}{\varepsilon_{DA}}\;, \tag{100}$$

which is also written as follows:

$$\frac{c_{0A}\cdot c_{0D}}{D}=\frac{c_{0A}\cdot c_{0D}\cdot d}{A_{exp}}=\frac{1}{K_c\cdot\varepsilon_{DA}}+\frac{1}{\varepsilon_{DA}}\cdot c_{0D}\;. \tag{100a}$$

The left-hand side of Eq. (100a) is plotted against c_{0D} and ε_{DA} is determined from the slope and $K_c \times \varepsilon_{DA}$ from the intercept with the ordinate.

According to (100), on the other hand, $c_{0A}\,d/A_e$ is plotted against $1/c_{0D}$: slope $= (K_c \times \varepsilon_{DA})^{-1}$; ordinate intercept: $(\varepsilon_{DA})^{-1}$.

As an example of the formation of an EDA complex, Fig. 58 shows the spectra of the *durol/chloranil equilibrium system* in CCl_4. It can be seen that the absorption spectrum at $\lambda = 481$ nm (corresponding to 20800 cm^{-1}) does not underlie the absorption of the acceptor (chloranil) and donor (durol) to any significant extent. That means that the requirement $\varepsilon_A = \varepsilon_D = 0$ applies here and to the remaining part of the absorption band toward longer wavelength (smaller wavenumbers).

The concentration of durol (donor) was varied and the acceptor concentration was kept constant in this example. This permits a graphical evaluation by Eq. (100) or, if we also want to evaluate at $\lambda < \lambda_{max}$ or $\tilde{\nu} > \tilde{\nu}_{max}$, by Eq. (99a) since the absorption of the acceptor in this region cannot be neglected. Figure 59 shows the plot from Eq. (100). A least squares calculation yields the equation of the best straight line as:

$$Y = 1.802\times10^{-4} + 3.873\times10^{-4}$$

with

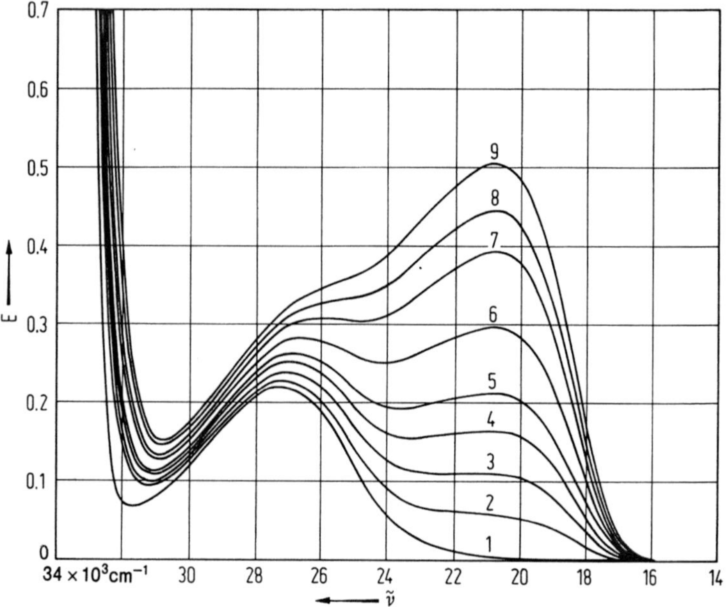

Fig. 58. EDA-complex formation in the system durol/chloranil; $c_{\text{chloranil}} = 9.69\times10^{-4}$ M; c_{durol} $1 = 1.033\times10^{-2} - 9 = 12.4\times10^{-2}$; ambient temperature; d = 1 cm

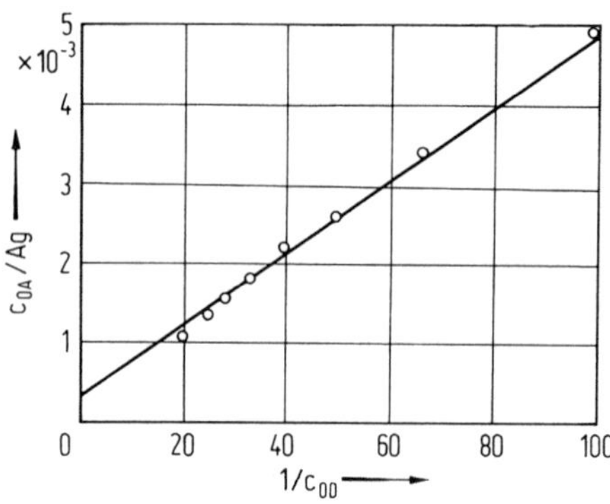

Fig. 59. Evaluation of the spectra in Fig. 58 according to Eq. (100)

$$Y = \frac{c_{0A}\cdot d}{A} \quad \text{and} \quad x = \frac{1}{c_{0D}}\,.$$

From the intercept we obtain $(\varepsilon_{DA})^{-1} = 3.873\times10^{-4}$ or $\varepsilon_{DA} = 2580\pm300$ [l mol^{-1} cm^{-1}].

From the slope we find $(K_c\varepsilon_{DA})^{-1} = 1.802\times10^{-4}$.

Hence, $K_c\varepsilon_{DA} = 5549$ and $K_c = 2.15\pm0.26$ [l mol^{-1}].

Figure 59 shows that errors of measurement have considerable influence on the determination of the intercept on the ordinate.

This evaluation contains another error because the *activity coefficients* have been neglected in the initial approximation. We require these coefficients when making a graphical evaluation in order to derive the relationships in Eqs. (98) to (100a) (See Briegleb [53] and Scott [54] for a discussion of the influence of activity effects).

All the relationships (99) to (100a) are suitable for graphical evaluation. As noted above, due to unavoidable errors of measurement there are occasionally considerable uncertainties in the value of K_c and ε_{DA}. Liptay [55] has developed a numerical method in which the measurement of different sets of concentration data are combined with the advantages of evaluation at different wavelengths; in the first instance, this provides information about the number of complexes formed.

Starting from equilibrium (94), we can generally specify the absorbance A_i or optical density D_i of a solution measured at wavelength i as follows ($A_i = D_i$ for d = 1 cm):

$$D_i = \varepsilon_{i,D_mA_n} \cdot c_{D_mA_n} - \varepsilon_{i.D} \cdot m\, c_{D_mA_n} - n\, \varepsilon_{i.A}\, \varepsilon_{D_mA_n} + \varepsilon_{i.D}\, c_{0D} + \varepsilon_{i.A}\, c_{0A} \ . \qquad (101)$$

If we bring the constant terms for every solution to the left-hand side we obtain

$$D_i' = D_i - \varepsilon_{i.D} \cdot c_{0D} - \varepsilon_{i.A}\, c_{0A} = (\varepsilon_{i.D_mA_n} - m\, \varepsilon_{i.D} - n\, \varepsilon_{i.A})\, c_{D_mA_n}$$

$$D_i' = \Delta\varepsilon_i \cdot c_{D_mA_n}$$

with

$$\Delta\varepsilon_i = \varepsilon_{i.D_mA_n} - m\, \varepsilon_{i.D} - n\, \varepsilon_{i.A} \ . \quad \text{[(see Eq. (96)]} \qquad (102)$$

However, the terms D_i' are a function of concentrations c_{0D} and c_{0A}. If the measurement of q solutions is combined with the measurement of p wavelengths the resulting values D_{iK}' can be summarized as a matrix:

$$D_{ik}' = \begin{pmatrix} D_{11}' & D_{12}' & D_{13}' & \dots & D_{1q}' \\ D_{21}' & D_{22}' & D_{23}' & \dots & D_{2q}' \\ \dots & \dots & \dots & \dots & \dots \\ D_{p1}' & D_{p2}' & D_{p3}' & \dots & D_{pq}' \end{pmatrix} \ . \qquad (103)$$

Since it was assumed that there was only one equilibrium and only one complex D_mA_n, in addition to components D and A, the matrix rank equals one. Therefore, all columns and rows must be proportional to each other.

This does not apply if other association equilibria are present. If there are two equilibria the matrix rank equals two.

When the matrix (103) is normalized to a specific frequency m, i.e. we set

$$D_{mk} = 1 \text{ for all solutions k} \qquad (104)$$

and the values

$$\varrho_{ik} = \frac{D_{ik}}{D_{mk}} \tag{105}$$

are calculated for all solutions k and wavelengths i. A new matrix results:

$$(\varrho_{ik}) = \begin{pmatrix} \varrho_{11} & \varrho_{12} & \cdots & \varrho_{1q} \\ \varrho_{21} & \varrho_{22} & \cdots & \varrho_{2q} \\ \cdots & \cdots & \cdots & \cdots \\ \varrho_{p1} & \varrho_{p2} & \cdots & \varrho_{pq} \end{pmatrix} . \tag{106}$$

If the matrix rank equals one all columns in (106) must be equal within the experimental accuracy, i.e.

$$\varrho_{11} = \varrho_{12} = \ldots = \varrho_{iq} . \tag{107}$$

Conversely, if the experimental values conform with the relationship (107) there is, in general, only one complex in the solution and its absorption differs from that of the components in the wavelength region under investigation.

We use *naphthalene/tetracyanoethylene* (TCE) as a numerical example of a *matrix analysis*. Figure 60 shows firstly the spectra obtained for different donor concentrations at a constant acceptor concentration.

The spectrum of the complex shows two absorption bands of different intensity. The first maximum lies at

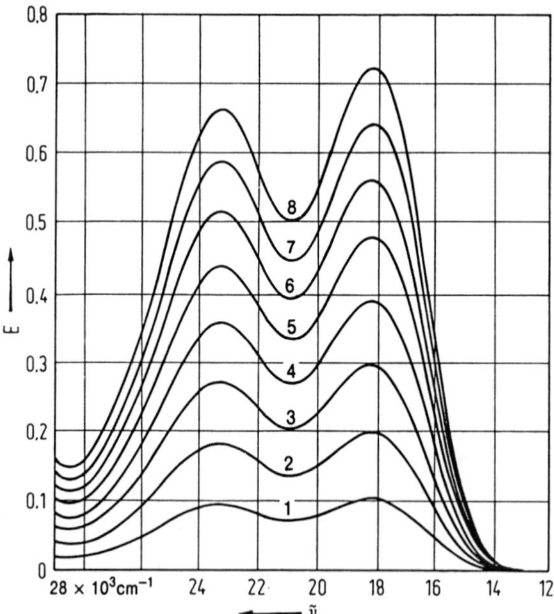

Fig. 60. EDA-complex formation naphthalene/tetracyanoethylene (TCE); see Table 18 for the concentration ratios; ambient temperature; d = 1 cm

Table 16. Experimental D'_{ik}, matrix (103); system: naphthalene/TCE

i	$\tilde{\nu}\,cm^{-1}$	D_{i1}	D_{i2}	D_{i3}	D_{i4}	D_{i5}	D_{i6}	D_{i7}	D_{i8}
1	16000	0.045	0.086	0.140	0.175	0.220	0.252	0.290	0.325
2	17000	0.082	0.160	0.240	0.315	0.397	0.455	0.525	0.592
3	18200	0.100	0.196	0.293	0.386	0.479	0.560	0.640	0.720
4	19000	0.092	0.182	0.272	0.358	0.442	0.520	0.593	0.670
5	20000	0.075	0.148	0.222	0.292	0.363	0.428	0.488	0.550
6	21000	0.070	0.135	0.201	0.266	0.330	0.388	0.443	0.501
7	22000	0.078	0.153	0.230	0.304	0.375	0.440	0.505	0.570
8	23400	0.092	0.180	0.270	0.353	0.436	0.512	0.586	0.660
9	24000	0.087	0.170	0.256	0.335	0.413	0.488	0.555	0.626

Table 17. ϱ_{ik} values, matrix (106) derived from matrix (103), Table 16

i	ϱ_{i1}	ϱ_{i2}	ϱ_{i3}	ϱ_{i4}	ϱ_{i5}	ϱ_{i6}	ϱ_{i7}	ϱ_{i8}
1	0.450	0.439	0.478	0.453	0.459	0.450	0.453	0.451
2	0.820	0.816	0.819	0.816	0.829	0.873	0.820	0.822
3	1	1	1	1	1	1	1	1
4	0.920	0.929	0.928	0.927	0.922	0.928	0.926	0.931
5	0.750	0.775	0.758	0.756	0.758	0.764	0.763	0.763
6	0.700	0.689	0.686	0.689	0.689	0.693	0.692	0.690
7	0.78	0.781	0.785	0.788	0.783	0.786	0.789	0.792
8	0.920	0.918	0.922	0.915	0.910	0.914	0.916	0.917
9	0.870	0.867	0.874	0.868	0.862	0.871	0.867	0.869

$\tilde{\nu}_{1,max} = 18\,400\ cm^{-1}$ (equivalent to 543 nm) and the second maximum at

$\tilde{\nu}_{2,max} = 23\,400\ cm^{-1}$ (equivalent to 427 nm).

Thus, the question is: are two EDA complexes formed or is the CT-absorption band of the EDA complex split? Table 16 shows the D_{ik} values which, at $d = 1$ cm, correspond to absorbances A_i measured at 9 wavelengths and in 8 solutions.

Table 17 shows the values of matrix (106). The ϱ_{ik} values are found to be constant for each row i. This means that the matrix (16) has a rank of one and that only *one* EDA complex is formed. Briegleb and Czekalla [56, 57] have explained the occurrence of double bands in this system and in other EDA complexes.

Of course, a statement that the matrix (103) is of rank one is exactly the same as the formulation of an absorbance-difference diagram (ΔA diagram), which we described as a graphical matrix rank method for spectrophotometric titration. Consequently, a ΔA diagram can be used to determine graphically whether only one complex is formed.

Here again, the Benesi-Hildebrand equation can be used for evaluation. Table 18 shows concentrations c_{0A} = const, c_{0D} = variable and the absor-

Table 18. Concentration c_{0D}, absorbances for $\lambda_1 = 543$ nm, $\lambda_2 = 427$ nm and calculated values for the evaluation of the naphthalene/TCE system as shown in Eqs. (98) and (100), $c_{0A} = 1.981 \times 10^{-3}$ mol/l

k	$c_{0D} \times 10^2$ [mol·l]	A_{543}	A_{427}	$\frac{c_{0A}}{A_{543}} \times 10^3$	$\frac{c_{0A}}{A_{427}} \times 10^3$	$\frac{c_{0A} \cdot c_{0D} \times 10^3}{(c_{0A} + c_{0D}) A_{543}}$
1	6.977	0.100	0.092	19.81	21.53	19.26
2	13.954	0.196	0.180	10.11	11.00	9.96
3	20.931	0.293	0.270	6.76	7.33	6.69
4	27.908	0.386	0.353	5.13	5.61	5.10
5	34.885	0.479	0.436	4.14	4.54	4.11
6	41.862	0.560	0.512	3.86	3.86	3.52
7	48.839	0.640	0.586	3.09	3.38	3.08
8	55.816	0.720	0.660	2.75	3.00	2.74

bance values at wavelengths $\lambda_1 = 543$ nm and $\lambda_2 = 427$ nm. Within the limits of error, an evaluation at $\lambda_1 = 543$ nm by Eq. (98) provides the same result as shown in Eq. (100). Regression analysis yields the best line in both cases as:

$$y = 1.364 \cdot 10^{-4} x + 2.541 \cdot 10^{-4}$$

with

$$y = \frac{c_{0A} \cdot c_{0D}}{c_{0A} + c_{0D}} \cdot \frac{1}{A_{543}} \quad \text{and} \quad x = \frac{1}{c_{0A} + c_{0D}} \; . \tag{98}$$

or

$$y = \frac{c_{0A}}{A_{543}} \quad \text{and} \quad x = \frac{1}{c_{0D}} \; . \tag{100}$$

Thus, the intercept on the ordinate lies at 2.541×10^{-4} with a mean deviation of $\pm 0.411 \times 10^{-4}$ and from this the extinction coefficient is $\varepsilon_{DA.543} = 3900 \pm 600$ [l mol^{-1} cm^{-1}).

Figure 61 shows the evaluation at both wavelength using Eq. (98).

By means of the method of least squares, the following values for the intercept on the ordinate are found:

$\lambda_1 = 543$ nm: $2.541 \times 10^{-4} \pm 0.411 \times 10^{-4}$
$\lambda_2 = 427$ nm: $3.232 \times 10^{-4} \pm 0.246 \times 10^{-4}$

and thus

$\varepsilon_{\lambda_1} = 3900 \pm 600$
$\varepsilon_{\lambda_2} = 3100 \pm 250$.

The great advantage of numerical evaluation using matrices (103) and (106) is that the accuracy of the results can be checked with one calculation combining regression analysis with error analysis. Liptay has described a

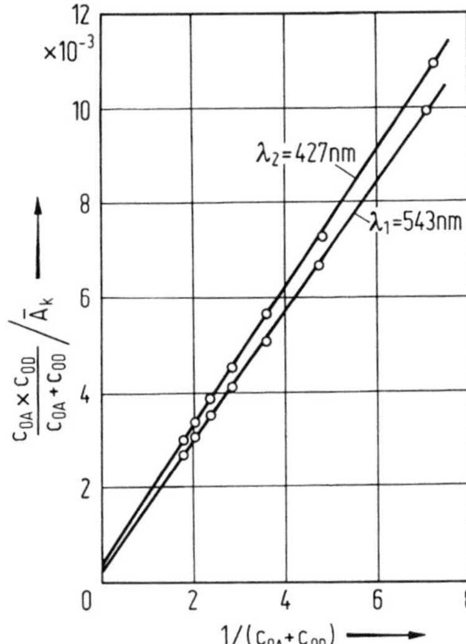

Fig. 61. Evaluation of the spectra in Fig. 93 using Eq. (98)

numerical evaluation of the Benesi-Hildbrand equations in detail [55]. By means of the definition (105) of matrix elements ϱ_{ik}, a calculation of the value D_{mk} at frequency m is possible for every D_{ik} at frequency i using

$$D_{mk} = \frac{D_{ik}}{\varrho_{ik}} . \qquad (108)$$

An arithmetical mean value D_k can be formed from the calculated D_{mk} values for every solution. The formation of a mean value does not provide a real improvement on the experimental results. Nevertheless, deviations from the mean value can be recognized and individual data points can be eliminated from the series of measurements and not considered in further calculations.

The numerical evaluation of the naphthalene /1,3,5-trinitrobenzene and benzene/1,3,5-trinitrobenzene systems [58] has been discussed [50]. Briegleb and Czekalla have investigated EDA-complex formation with chloranil as acceptor [59]. A detailed examination of errors is included in the description of the numerical evaluation [53, 55], and the accuracy of the experimental results in relation to the concentration (mole fraction or molarity) is also considered.

The question of the number of EDA complexes formed can be answered without difficulty by means of a numerical or graphical factor analysis. However, the question of *the stoichiometry of a complex* remains to be clarified. Here a method which was first described in 1910 and 1912 [60, 61] and which is called "Job's method" after its author [62] can be used.

Under the condition that the total concentration $C_0 = c_{0D} + c_{0A}$ is constant whilst concentrations c_{0D} and c_{0A} are *variable* then Eq. (94) still applies. If we make $c_{0D} = x$ and substitute $c_{0D} = c_0 - x$ we can write

$$K_c = \frac{c_{D_mA_n}}{(x - m c_{D_mA_n})^m (c_0 - x - n c_{D_mA_n})^n} \ . \tag{109}$$

The condition for a maximum concentration of complex D_mA_n, $c_{D_mA_n}$, as a function of the ratio of the different components $0 \le x \le 1$ is

$$\frac{d c_{D_mA_n}}{dx} = 0 \ . \tag{110}$$

Initially, the following results from (109):

$$c_{D_mA_n} = K_c \cdot (x - m c_{D_mA_n})^m \cdot (c_0 - x - n c_{D_mA_n})^n \ .$$

It follows that

$$m(x - m c_{D_mA_n})^{m-1} (c_0 - x - n c_{D_mA_n})^n = n(x - m c_{D_mA_n})^m (c_0 - x - n c_{D_mA_n})^{n-1} \tag{111}$$

and from this:

$$m(c_0 - x) = nx$$

or

$$m \cdot c_{0A} = n \cdot c_{0D} \ .$$

This method of continuous variation according to Job [60] permits the determination of the ratio m/n of the stoichiometric coefficients. However, it is reliable only if a *complex* is formed. The method can only be applied to coupled equilibria for favorable combinations of equilibrium constants and extinction coefficients [53]. In practice the differences between measured extinctions and those calculated on the basis that no complex is formed are plotted against composition $x = c_{0D}$. A curve with an extremum is obtained. According to our discussions this means that dD_i/dc_{0D} of Eq. (102) must assume a maximum or minimum value. See Schläfer for an explanation of this requirement [63].

6.3.3 Metal Complexes

During the discussion of EDA complexes it was assumed that *one* stoichiometrically defined complex is formed in an equilibrium. Especially in the case of the formation of *metal complexes*, this assumption does not often apply. Therefore, we have to take into account the possibility of several equilibria which are not independent of each other.

If we denote the metal ion by M and the ligand by L we can formulate the following equilibria:

$$M + L \overset{K_1}{\rightleftharpoons} ML \ , \quad K_1 = \frac{[ML]}{[M][L]} \ . \tag{112a}$$

$$ML+L \overset{K_2}{\rightleftharpoons} ML_2 \ , \quad K_2=\frac{[ML_2]}{[ML][L]} \ , \quad K_1\cdot K_2=\frac{[ML_2]}{[M][L]^2} \ . \tag{112b}$$

$$ML_{n-1}+L \overset{K_n}{\rightleftharpoons} ML_n \ , \quad K_n\frac{[ML_n]}{[ML_{n-1}][L]} \ , \quad K_1\cdot K_2\dots K_n=\frac{[ML_n]}{[M][L]^n} \ . \tag{112c}$$

The constants K_n are called *individual stability constants.* The product of the stability constants $K_1 \cdot K_2 \dots K_n = \bar{K}_n$ is called the *formation constant.* Since the formation constants of the intermediate stages of a complex are needed in many investigations it has proved to be appropriate to introduce the notation "β".

Thus the following relationships hold

$$\beta_1 = K_1 \ , \quad \beta_2 = K_1 \cdot K_2 \ , \quad \beta_n = K_1 \cdot K_2 \dots K_n \ , \tag{113}$$

and the constant β_n applies to the equilibrium (112c)

$$M+nL \rightleftharpoons ML_n \ , \quad \beta_n=\frac{[ML_n]}{[M][L]^n} \ .$$

See Sect. 4.1.1 for the significance of stability constants.

A determination of β by means of UV-VIS spectroscopy assumes that there is an appropriately defined complex in respect of its stoichiometric composition and that it is formed *exclusively.* Otherwise the stability range of a specific intermediate stage of a complex must be so large that there is no overlap with other intermediate stages. If this is the case then the equilibrium composition can be determined by means of a multicomponent analysis, if the extinction coefficients are known for the pure components associated with the formation equilibrium.

Frequently, the situation arises where the metal cation and/or ligand does not absorb in the spectral region in which a metal complex absorbs; this happens with many inorganic anions used as ligands. It does not always apply for organic ligands since their absorption spectra can overlap to a greater or lesser extent with those of the complexes formed.

Thus, three questions must always be answered when investigating complex-forming equilibria:

1. How many complexes are formed?
2. What types of stoichiometry do they have?
3. How large are the formation constants β_n?

It can easily be overlooked that these questions are more difficult to answer when more intermediate stages of a complex occur within small stability ranges. In the graphical representation, the curves defining the stability ranges correspond to the S-shaped titration curves. Whereas there we plotted against the pH-value, here we plot against $-\log [L] \equiv pL$ [68].

It is often appropriate to formulate the equilibrium (112c) as a dissociation equilibrium. If there is a sufficiently large interval between the following complex-forming stage or dissociation stage it can be assumed that the following applies:

$$c_{0M} = c_n + c_{n-1}$$

where $c_n = [ML_n]$, $c_{n-1} = [ML_{n-1}]$ and c_{0M} = the given concentration of the metal ion.

Thus, there results for the measured absorbance A_e:

$$A_e = \varepsilon_n \cdot c_{0M} \cdot d - \varepsilon_n c_{n-1} d + \varepsilon_{n-1} c_{n-1} \cdot d \ . \tag{114}$$

With the requirement that $c_{0M} = c_n + c_{n-1}$, expressions are obtained for equilibrium concentrations c_n and c_{n-1} and thus for the dissociation constant of the step $ML_n = ML_{n-1} + L$, analogous to the procedure for protolytic equilibria (see Eqs. (73) to (77), Sect. 6.2):

$$K'_n = \frac{A_e - A_{0n}}{A_{0,n-1} - A_e} \cdot [L] = \frac{\varepsilon'_e - \varepsilon_n}{\varepsilon_{n-1} - \varepsilon'_e} \cdot [L] \tag{115}$$

where $A_{0n} = \varepsilon_n \times c_{0n} d$ and $A_{0,n-1} = \varepsilon_{n-1} \times C_{0M} \times d$; ε'_e is again the *variable extinction coefficient* calculated from the measured absorbance A_e and the constant concentration of metal ions c_{0M} (see Sect. 6.2).

By solving (115) for ε'_e, Eq. (116) is obtained, analogous to Eq. (81), which describes the S-shaped dissociation curves:

$$\varepsilon'_e = \frac{\varepsilon_{n-1} \cdot 10^{-pK'_n} + \varepsilon_n \cdot 10^{-pL}}{10^{-pL} + 10^{-pK'_n}} \ . \tag{116}$$

If $pL = pK'_n$ then $\varepsilon'_e = \frac{\varepsilon_{n-1} - \varepsilon_n}{2}$ results or, if the complex $n-1$ does not absorb (i.e. $\varepsilon_{n-1} = 0$), $\varepsilon_e = \varepsilon_n/2$ and thus the pL-value corresponds to the pK_n-value at the inflexion point of the S-shaped curve; and $pK_n = -pK'_n$. If the logarithm of Eq. (115) is taken the Henderson-Hasselbach equation is again obtained (78a):

$$\log \frac{\varepsilon'_e - \varepsilon_n}{\varepsilon_{n-1} - \varepsilon'_e} = pL - pK'_n \ . \tag{117}$$

If the left-hand side is plotted against $pL = -\log L$ (see Sect. 6.2) a graphical evaluation of this complex-formation stage and the determination of pK_n can be carried out.

It can be seen that in these simple cases this is a titration system. However, in complicated cases we can speak of multi-stage complex-forming systems or *dissociation equilibria* which overlap to a greater or lesser extent. Consequently, all the evaluation methods discussed in Sect. 6.2 can also be applied to complex-forming equilibria.

When using the evaluation shown in Eq. (117) we must know the extinction coefficients ε_n for ML_n and ε_{n-1} for ML_{n-1}. If ε_{n-1} equals zero in the region where complex ML_n absorbs ε_n can be determined directly by using a great excess of metal ions or ligands. However, this method is only reliable if the stability constant is very large.

In principle, with small stability constants an evaluation can be made using the Benesi-Hildebrand equation, see Eq. (100), to obtain the extinction coefficient and the K_n-value by means of a graphical evaluation (see examples in Sect. 6.2.2).

The question concerning the number of complex-forming stages over a wide range of variable ligand concentrations can be answered using a graphical matrix rank analysis of the spectroscopic data.

When recording spectra at a constant metal-ion concentration as a function of a variable ligand concentration we obtain a host of overlapping spectra (similar to the spectra shown in Sect. 6.2.2). If only one complex is formed in the concentration range under investigation then, for a given wavelength, the absorbance-difference diagram (ΔA diagram) must result in a straight line, see Eq. (86), which passes through the origin.

If deviations from linearity occur this indicates that more than one complex is being formed in the case of two complex-forming stages one following the other, the ΔAQ diagram (absorbance-difference-quotient diagram) must result in a linear relation (see Sect. 7.2). We referred to this evaluation when discussing multi-stage photometric titration systems.

The second question to be answered concerns the *stoichiometry of the metal complex*. If a complex of the general stoichiometry ML_n is formed then Eq. (112c) applies. If the equilibrium concentrations ML_n and M are expressed as absorbances and the logarithm is taken of this expression, there results:

$$\log K_n = \log \frac{A_e - A_M}{A_n - A_e} - n \log [L] \tag{118}$$

or

$$\log \frac{\varepsilon'_e - \varepsilon_M}{\varepsilon_n - \varepsilon'_e} = n\,pL - pK \;. \tag{118a}$$

Thus, if $\log \frac{\varepsilon'_e - \varepsilon_M}{\varepsilon_n - \varepsilon'_e}$ is plotted against pL a straight line is obtained, the slope of which provides the stoichiometric coefficient n, i.e. the number of ligands. In Eqs. (118) and (118a) account was taken of the fact that the metal ion, but not the ligand, absorbs in the spectral region under investigation.

If a complex has the composition $M_m L_n$ we can use the method of continuous variation discussed for EDA complexes. It yields the ratio of the coefficients.

Vosburgh and Cooper [71] have extended this method and applied it to the case of the formation of several complexes. Katzin and Gebert [72] have discussed the formation of three complexes.

Woldby has detailed the scope of validity of the method of continuous variation [73]. The formation of a single complex and the stepwise formation of different types of complexes must always be differentiated. Schläfer has given a detailed description of the different methods [63]. If the stoichiometry is known and it is assumed that only one complex is formed the continuous variation method can be used to determine the stability constants.

Schläfer [63] and Beck [74] have compiled advanced texts dealing with spectrophotometric or photometric determination of the formation constants of metal complexes.

Particular difficulties occur when determining stability constants if a ligand is capable of reacting protolytically because the pH-dependence of the complex-forming equilibrium is then involved. Polster [75] has shown that photometric titration with acids or bases can be used for determining the formation constants β. By using the absorbance diagrams (A-diagrams: A_{λ_1}(pH) plotted against A_{λ_2}(pH)) it is possible to determine concentrations c_L, c_M and C_{LM} for every pH-value [22] from which the stability constant can be calculated.

To conclude this chapter, it is worth noting that the thermodynamic data for the equilibria discussed here can be determined by including the dependence on temperature.

References

1. Henderson LJ, Tannenbaum M (1932) Blut, seine Pathologie und Physiologie. Steinkopff, Dresden
2. Rosenblatt DH (1954) J Phys Chem 58:40
3. King EJ (1965) Acid-Base Equilibria. Pergamon Press, London
4. Rabenstein DL, Greenburg MS, Erans CA (1977) Biochemistry 16:977
5. Kortüm G, Vogel W, Andrussow K (1962) Pure Appl Chem 1:190
6. Kortüm G (1972) Lehrbuch der Elektrochemie. Verlag Chemie, Weinheim
7. Blume R, Lachmann H, Mauser H, Schneider F (1974) Z Naturforsch 29b:500
8. Lachmann H, Polster J (1982) Spektrometrische Titrationen. Vieweg, Wiesbaden
9. Lachmann H (1982) Habilitationsschrift. Univers Tübingen
10. Ebel S, Parzefall W (1975) Experimentelle Einführung in die Potentiometrie. Verlag Chemie, Weinheim. In: Ebel S, Surmann P (1980) Ullmanns Encyclopädie der Technischen Chemie, Bd 5. Chemie, Weinheim, S 651–684
11. Tubbs CF (1954) Anal Chem 26:1670
12. Perkampus H-H, Prescher G (1968) Ber Bunsenges Physik Chemie 72:429
13. Judson CM, Kilpatrick M (1949) J Amer Chem Soc 71:3110
14. Kortüm G, Shih C (1977) Ber Bunsenges Physikal Chem 81:44
15. Blume R, Polster J (1974) Z Naturforsch 29b:794
16. Perkampus H-H, Rössel Th (1956) Z Elektrochemie. Ber Bunsenges Physikal Chem 60:1102
17. Perkampus H-H, Rössel Th (1958) ibid 62:94
18. Polster J (1975) Z Physikal Chem NF 97:55
19. Polster J (1977) ibid 104:49
20. Polster J (1975) Fresenius Z Anal Chem 276:353
21. Göbber F, Polster J (1976) Anal Chem 48:1546

22. Blume R, Lachmann H, Polster J (1975) Z Naturforsch 30b, 263
23. Göbber F, Lachmann H (1978) Hoppe-Seylers Z physiol Chem 359:269
24. Nagano K, Metzler DE (1967) J Amer Chem Soc 89:2891
25. Sullivan JC (1959) Acta Chem Scand 13:2023
26. Sillen LG (1962) 16:159, 173; (1964) ibid 18:1085
27. Scheibe G (1938) Kolloid Z 82:2
28. Scheibe G (1939) Angew Chem 52:631; (1948) Z Elektrochemie 52:283 (1948)
29. Kempter H, Mecke R (1940) Z physik Chem B46, 229
30. Coggeshall ND, Saier EL (1951) J Amer Chem Soc 73:5414; Saarla-Mathot L (1953) Trans Faraday Soc 49:8;
Coburn WC, Grünwald E (1958) J Amer Chem Soc 80:1318; Frank HL, Wen WY (1957) Disc Faraday Soc 24:133; Tucker E, Becker ED (1973) J Physic Chem 77:1783
31. Perkampus H-H, Kerim F (1968) Spectrochim Acta 24A:2071
32. Perkampus H-H, Juffernbruch J (1980) ibid 36A:485
33. Geiseler G, Mehmert E (1968) ibid 24A:943
34. Jaffé HH, Orchin M (1982) Theory and Application of Ultraviolet-Spectroscopy. Wiley, New York London
35. Murrell JN (1967) Elektronenspektren organischer Moleküle, Bd 250/250a, B. I. Hochschultaschenbücher. Bibliograph Inst, Mannheim
36. West W (1968) Chemical Applications of Spectroscopy. Wiley, New York London
37. Burawoy A (1939) J Chem Soc 1177
38. Coppens G, Gillet C, Nasidsky J, van der Donkt E (1962) Spectrochim Acta 18:1441
39. Nasielsky J, van der Donkt E (1963) ibid 19:1989
40. Perkampus H-H, Baucke F (1968) DMS-UV-Atlas. Perkampus H-H, Sandemann I, Timmons J (Hrsg) (1968) Butterworth, London; Verlag Chemie, Weinheim, Vol IV, Spektren H 18/7 und H 18/8
41. Perkampus H-H, Baucke F (1961) Z Elektrochem Ber Bunsenges Physik Chem 65:699
42. Brealey GJ, Kasha MJ (1955) J Amer Chem Soc 77:4462
43. Lippert E (1959) Der Einfluß von Wasserstoffbrücken auf Elektronenspektren. In: Hadzi D, Thompson HW (eds) Hydrogen Bonding. Pergamon Press, New York, London, Paris, Los Angeles
44. Pimentel GC (1957) J Amer Chem Soc 79:3323
45. Juffernbruch J, Perkampus H-H (1983) Spectrochim Acta 39A:905
46. Juffernbruch J (1982) Dissertat Univers Düsseldorf
47. Juffernbruch J, Perkampus H-H (1983) Spectrochim Acta A 39A:1093
48. Juffernbruch J, Perkampus H-H (1983) ibid 39A:1097
49. Seguin JP, Jadjo L, Uzan R, Doncet JP (1981) ibid 37A:205
50. Briegleb G (1961) Elektronen-Donator-Acceptor-Komplexe. Springer, Berlin Göttingen Heidelberg
51. Ketelaar ZAA, van de Stolpe C, Gondsmit A, Dzcubas W (1952) Rec Trav Chim 71:1104
52. Benesi HA, Hildebrand JH (1949) J Am Chem Soc 71:2703
53. Briegleb G, cit [50], Kap XII
54. Scott RL (1956) Rec Trav Chim 75:787
55. Liptay W (1961) Z Elektrochem 65:375
56. Briegleb G, Czekalla J (1960) Angew Chem 72:401
57. Briegleb G, Czekalla J, Reuss G (1961) Z physik Chem NF 30:316, 333
58. Briegleb G, Czekalla J (1955) Z Elektrochem 59:184
59. Briegleb G, Czekalla J (1954) Z Elektrochem 58:249
60. Ostromisslewsky I (1910) J Russ Phys Chem Ges 42:1932
61. Denison RB (1912) Trans Faraday Soc 8:20, 35
62. Job P (1928) Ann Chim Phys France 9:113; (1925) Compt Rend hebd Acad Sci France 180:928
63. Schläfer HL (1961) Komplexbildung in Lösung. Springer Verlag, Berlin Göttingen Heidelberg
64. Forster R (1969) Organic Charge-Transfer-Complexes. Academic Press, London New York San Francisco

65. Mulliken RS, Person WB (1969) Molecular Complexes. Wiley, New York London Sidney Toronto
66. Gur'yanova EN, Gol'dshtein IP, Romm IP (1975) Donor-Acceptor-Bond. Wiley, New York Toronto
67. Forster R (1975) Molecular Association, Vol I. Academic Press, London New York San Francisco
68. Schwarzenbach G (1961) Adv Inorg Radiochem 3:257
69. Mauser H (1968) Z Naturforsch 23b:1025
70. Mauser H (1974) Formale Kinetik. Bertelsmann Universitätsverlag, Düsseldorf
71. Vosburgh WC, Cooper RG (1941) J Am Chem Soc 63:437
72. Katzin LI, Gebert E (1950) ibid 72:5455
73. Woldbye F (1955) Acta Chem Scand 9:299
74. Beck MT (1970) Chemistry of Complex Equilibria. Van Nostrand Reinhold Co, London
75. Polster J (1976) Z Naturforsch 31b:1621

7 Investigation of the Kinetics of Chemical Reactions

The investigation of the kinetics of chemical reactions by monitoring the absorbance with respect to time has already been mentioned in the introduction to chapter 6. In this way the order of a reaction, the velocity constant and, by taking account of the temperature dependence, the energy of activation can be determined.

The absorbance $A_{\tilde{\nu}}$ or A_λ is used as one of the variables proportional to concentration in order to follow the change in concentration of reactants and products with respect to time. This has the advantage that the system under investigation is not disturbed, provided that it is photochemically stable.

In the routine application of UV-VIS spectroscopy using recording dual-beam spectrophotometers, the duration of the measurements can extend from minutes to several hours, and under certain circumstances, to days. Basically, we are dealing with slow reactions. Special methods must employed when investigating *fast reactions.*

7.1 Fundamental Equations of Kinetics

7.1.1 Introduction of Absorbance as a Measurement Parameter

The basics of chemical kinetics can be found in the textbooks of physical chemistry and chemical kinetics [1–5]. For that reason, only the relationships involving the absorbance as a parameter will be considered here.

If we consider a general reaction of form:

$$\nu_a a + \nu_b b + \nu_c c \rightarrow \nu_p p \ , \quad (\nu_i\text{: stoichiometric coefficient of component i}) \tag{119}$$

then, for an absorbance measured at a specific wavelength "λ" and at time "t" of the reaction:

$$A_\lambda = a\varepsilon_{a,\lambda} d + b\varepsilon_{b,\lambda} d + c\varepsilon_{c,\lambda} d + p\varepsilon_p d \ . \tag{120}$$

a, b, c and p are the concentrations at time t. By introducing the transformation variable "x" we obtain for the appropriate concentrations:

$$a = a_0 + \nu_a x \ , \quad b = b_0 + \nu_b x \ , \quad c = c_0 + \nu_c x \ , \quad p = p_0 + \nu_p x \ .$$

a_0, b_0 and c_0 are the initial concentrations when $t = 0$. p_0 then equals zero, and from Eq. (120) it follows that:

$$A_\lambda = (a_0 + \nu_a x)\varepsilon_{a,\lambda} d + (b_0 + \nu_b x)\varepsilon_{b,\lambda} d + (c_0 + \nu_c x)\varepsilon_{c,\lambda} d + \nu_p x \varepsilon_{p,\lambda} \; . \qquad (121)$$

Removing the brackets and rearranging gives the expression:

$$A_\lambda = a_0 \varepsilon_{a,\lambda} d + b_0 \varepsilon_{b,\lambda} d + c_0 \varepsilon_{c,\lambda} d + d\,[\nu_p \varepsilon_{p,\lambda} + \nu_a \varepsilon_{a,\lambda} + \nu_b \varepsilon_{b,\lambda} + \nu_c \varepsilon_{c,\lambda}]\, x \; .$$

When $t = 0$, $x = 0$ from which we obtain the absorbance $A_{\lambda,0}$. With the definition:

$$q_\lambda = (\nu_p \varepsilon_{p,\lambda} + \nu_a \varepsilon_{a,\lambda} + \nu_b \varepsilon_{b,\lambda} + \nu_c \varepsilon_{c,\lambda})\, d \qquad (122)$$

or

$$q_\lambda = d \sum_i \nu_i \varepsilon_{i,\lambda}$$

it follows that:

$$A_\lambda - A_{\lambda,0} = q_\lambda \cdot x \; . \qquad (123)$$

Differentiation of Eq. (123) with respect to time provides a simple correlation between the change with time of the transformation variable x and the absorbance A_λ:

$$\frac{1}{q_\lambda} \cdot \frac{d(A_\lambda - A_{\lambda,0})}{dt} = \frac{1}{q_\lambda} \cdot \frac{dA_\lambda}{dt} = \frac{dx}{dt} \; . \qquad (124)$$

From Eq. (123) it also follows that:

$$x = \frac{A_\lambda - A_{\lambda,0}}{q_\lambda} \; . \qquad (123\,a)$$

For *simple* and *uniform* reactions* the rate laws can be formulated in terms of the absorbance by means of Eq. (124). This can be carried out explicitly on a 1st order reaction of type $a \rightarrow p$, see also [5]. If $\nu_a = -1$ and $\nu_p = 1$ and with k_1 as the rate constant the rate law reads:

$$\frac{dx}{dt} = k_1 (a_0 - x) \; .$$

---

* A "simple" reaction is one for which a simple rate law applies. A "uniform" reaction is one where no complicating features such as back-reactions, equilibria or branching occur

From Eqs. (123) and (124) it follows that:

$$\frac{dA_\lambda}{dt} = q_\lambda k_1 \left(a_0 - \frac{A_\lambda - A_{\lambda,0}}{q_\lambda} \right) = k_1 (a_0 q_\lambda - A_\lambda + A_{\lambda,0}) \ .$$

From Eq. (122), $q_\lambda = d(\varepsilon_{p,\lambda} - \varepsilon_{a,\lambda})$ which, when inserted in the above equation, gives:

$$\frac{dA_\lambda}{dt} = k_1 (a_0 \varepsilon_{p,\lambda} d - a_0 \varepsilon_{a,\lambda} d - A_\lambda + A_{\lambda,0}) \ , \qquad \frac{dA_\lambda}{dt} = k_1 (A_{\lambda,\infty} - A_\lambda) \ . \quad (125)$$

Because $A_{\lambda,0} = a_0 \varepsilon_{a,\lambda} d$ and $A_{\lambda,\infty} = a_0 \varepsilon_{p,\lambda} d$, i.e. the reactant has been completely converted to product p at $t \to \infty$: $a_0 = p_\infty$.

Integration of Eq. (125) with the condition that $A_\lambda = A_{\lambda,0}$ at time $t = 0$ provides:

$$\ln \frac{A_{\lambda,\infty} - A_\lambda}{A_{\lambda,\infty} - A_{\lambda,0}} = -k_1 t \ . \qquad (125\,a)$$

If the product does not absorb at wavelength λ, then $\varepsilon_{p,\lambda} = 0$ and $A_{\lambda\infty} = 0$, i.e. only the decrease of a is monitored at wavelength λ and it follows from Eq. (125) that

$$\ln \frac{A_\lambda}{A_{\lambda,0}} = -k_1 t \ ; \quad A_\lambda = A_{\lambda,0} e^{-k_1 t} \ . \qquad (125\,b)$$

In the case of $\varepsilon_{a,\lambda} = 0$, $A_{\lambda,0} = 0$ it follows analogously that:

$$\ln \frac{A_{\lambda,\infty} - A_\lambda}{A_{\lambda,\infty}} = -k_1 t \ ; \quad A_\lambda = A_{\lambda,\infty} (1 - e^{-k_1 t}) \ . \qquad (125\,c)$$

1.2 Classification of Other Types of Reaction

1.2.1 2nd Order Reactions

Reaction type a)	$2a \to p$
Stoichiometric coefficients	$\nu_a = -2 \ ; \quad \nu_p = 1$
$q_\lambda = d \sum \nu_i \cdot \varepsilon_{i,\lambda}$	$d(\varepsilon_{p,\lambda} - 2\varepsilon_{a,\lambda})$
$A_{\lambda,0} \ ; \quad A_{\lambda,\infty}$	$A_{\lambda,0} = a_0 \varepsilon_{a,\lambda} d \ ; \quad A_{\lambda,\infty} = \frac{a_0}{2} \varepsilon_{p,\lambda} d$

Rate law, general $\frac{dx}{dt} = k_2(a_0 - 2x)^2$

Rate law, absorbance $\frac{dA_\lambda}{dt} = \frac{4k_2}{q_\lambda}(A_{\lambda,\infty} - A_\lambda)^2$

Integrated rate law
$$\frac{1}{A_{\lambda,\infty} - A_\lambda} - \frac{1}{A_{\lambda,\infty} - A_{\lambda,0}} = \frac{4k_2}{q_\lambda} \cdot t \qquad (126)$$

$A_{\lambda,\infty} - A_{\lambda,0}$ $\frac{a_0}{2} q_\lambda$

Transformation of Eq. (126)
$$\frac{A_\lambda - A_{\lambda,0}}{t} = 2k_2 a_0 A_{\lambda,\infty} - 2k_2 a_0 A_\lambda \qquad (126a)$$

Evaluation $(A_\lambda - A_{\lambda,0})/t$ against A_λ

Special cases

1) $\varepsilon_{a,\lambda} \neq 0$; $\varepsilon_{p,\lambda} = 0$; $A_{\lambda,\infty} = 0$
$q_\lambda = -2\varepsilon_{a,\lambda} d$

$$\frac{1}{A_\lambda} - \frac{1}{A_{\lambda,0}} = \frac{2k_2}{\varepsilon_{a,\lambda} d} \cdot t \qquad (126b)$$

$$\frac{A_{\lambda,0} - A_\lambda}{t} = 2k_2 a_0 \cdot A_\lambda \qquad (126c)$$

2) $\varepsilon_{a,\lambda} = 0$; $\varepsilon_{p,\lambda} \neq 0$; $A_{\lambda,0} = 0$;

$$A_{\lambda,\infty} = \frac{a_0}{2} \varepsilon_{p,\lambda} \cdot d \; ; \quad q_\lambda = \varepsilon_{p,\lambda} \cdot d$$

$$\frac{A_\lambda}{A_{\lambda,\infty} - A_\lambda} = 2k_2 a_0 \cdot t \qquad (126d)$$

$$\frac{A_\lambda}{t} = 2k_2 a_0 A_{\lambda,\infty} - 2k_2 a_0 A_\lambda \qquad (126e)$$

Reaction type b) $a + b \rightarrow p \quad a_0 = b_0$

Stoichiometric coefficients $\nu_a = \nu_b = -1$; $\nu_p = 1$

$q_\lambda = d \sum \nu_i \varepsilon_{i,\lambda}$ $d(\varepsilon_{p,\lambda} - \varepsilon_{a,\lambda} - \varepsilon_{b,\lambda})$

$A_{\lambda,0}$; $A_{\lambda,\infty}$ $A_{\lambda,0} = a_0 \varepsilon_{a,\lambda} d + a_0 \varepsilon_{b,\lambda} d$; $A_{\lambda,\infty} = a_0 \varepsilon_{p,\lambda} d$

Rate law, general $\frac{dx}{dt} = k_2(a_0 - x)^2$

Rate law, absorbance $$\frac{dA_\lambda}{dt} = \frac{k_2}{q_\lambda}(A_{\lambda,\infty} - A_\lambda)^2 \qquad (127)$$

Integrated rate law $$\frac{1}{A_{\lambda,\infty} - A_\lambda} - \frac{1}{A_{\lambda,\infty} - A_{\lambda,0}} = \frac{k_2}{q} t \qquad (127\,a)$$

$A_{\lambda,\infty} - A_{\lambda,0}$ $\quad a_0(\varepsilon_{p,\lambda} - \varepsilon_{a,\lambda} - \varepsilon_{b,\lambda}) \cdot d = a_0 q_\lambda$

Transformation of (127a) $$\frac{A_\lambda - A_{\lambda,0}}{t} = k_2 a_0 A_{\lambda,\infty} - k_2 a_0 A_\lambda \qquad (127\,b)$$

Evaluation (example see [6]) $\quad (A_\lambda - A_{\lambda,0})/t$ against A_λ ;

Special cases $\quad$ analogous to reaction type a)

Reaction type c) $\quad a + b \rightarrow p$; $a_0 \neq b_0$; $b_0 > a_0$

Stoichiometric coefficients $\quad \nu_a = \nu_b = -1$; $\nu_p = 1$

$q_\lambda = d \sum \nu_i \varepsilon_{i,\lambda}$ $\quad d(\varepsilon_{c,\lambda} - \varepsilon_{a,\lambda} - \varepsilon_{b,\lambda})$

$A_{\lambda,0}$; $A_{\lambda,\infty}$ $\quad A_{\lambda,0} = a_0 \varepsilon_{a,\lambda} d + b_0 \varepsilon_{b,\lambda} d$;

$A_{\lambda,\infty} = [a_0 \varepsilon_{p,\lambda} + (b_0 - a_0) \varepsilon_{b,a}] d$

Rate law, general $$\frac{dx}{dt} = k_2 (a_0 - x)(b_0 - x)$$

Rate law, absorbance $$\frac{dA_\lambda}{dt} = \frac{k_2}{q_\lambda}(A_{\lambda,\infty} - A_\lambda)(A'_{\lambda,\infty} - A_\lambda)$$

$A'_{\lambda,\infty}$ (calculated parameter [5]) $\quad A'_{\lambda,\infty} = [b_0 \varepsilon_{p,\lambda} - (b_0 - a_0) \varepsilon_{a,\lambda}] d$

integrated (partial fraction factorization) [7, 8]
$$\ln \frac{A'_{\lambda,\infty} - A_\lambda}{A_{\lambda,\infty} - A_\lambda} = \ln \frac{A'_{\lambda,\infty} - A_{\lambda,0}}{A_{\lambda,\infty} - A_{\lambda,0}} + (A'_{\lambda,\infty} - A_{\lambda,\infty}) \frac{k_2}{q_\lambda} t \qquad (128)$$

.2.2 3rd Order Reactions

Reaction type a) $\quad 3a \rightarrow p$

Stoichiometric coefficients $\quad \nu_a = -3$; $\nu_p = 1$

$q_\lambda = d \sum \nu_i \varepsilon_{i,\lambda}$ $\quad d(\varepsilon_{p,\lambda} - 3\varepsilon_{a,\lambda})$

$A_{\lambda,0}$; $A_{\lambda,\infty}$ $\quad A_{\lambda,0} = a_0 \cdot \varepsilon_{a,\lambda} d$; $A_{\lambda,\infty} = \frac{a_0}{3} \varepsilon_{p,\lambda} d$

Rate law, general $\frac{dx}{dt} = k_3(a_0 - 3x)^3$

Rate law, absorbance $\frac{dA_\lambda}{dt} = \frac{27k_3}{q_\lambda^2}(A_{\lambda,\infty} - A_\lambda)^3$

Integrated rate law $$\frac{1}{(A_{\lambda,\infty} - A_\lambda)^2} - \frac{1}{(A_{\lambda,\infty} - A_{\lambda,0})^2} = \frac{54k_3}{q_\lambda^2} \cdot t \qquad (129)$$

$t = 0$; $A_\lambda = A_{\lambda,0}$

Special cases

1) $A_{\lambda,\infty} = 0$; $q_\lambda = -3\varepsilon_{a,\lambda} \cdot d$;

$A_{\lambda,0} = a_0 \varepsilon_{a,\lambda} \cdot d$;

$$\frac{1}{A_\lambda^2} - \frac{1}{A_{\lambda,0}^2} = \frac{54k_3}{q_\lambda^2} \cdot t \qquad (129\text{a})$$

or $$\frac{1}{\left(\frac{A_{\lambda,0}}{A_\lambda}\right)^2 - 1} = \frac{1}{6k_3 a_0^2} \cdot \frac{1}{t}$$

2) $A_{\lambda,0} = 0$; $q_\lambda = \varepsilon_{p,\lambda} d$; $A_{\lambda,\infty} = \frac{a_0}{3} \varepsilon_{p,\lambda} d$

after transformation via the reciprocal value of (129):

$$\frac{1}{2\left(\frac{A_\lambda}{A_{\lambda,\infty}}\right) - \left(\frac{A_\lambda}{A_{\lambda,\infty}}\right)^2} = \frac{1}{6k_3 a_0^2} \cdot \frac{1}{t} + 1 \qquad (129\text{b})$$

Reaction type b) $2a + b \rightarrow p$; $a_0 = 2b_0$

Stoichiometric coefficients $\nu_a = -2$; $\nu_b = -1$; $\nu_p = 1$

$q_\lambda = d \sum \nu_i \varepsilon_{i,\lambda}$ $q_\lambda = d(\varepsilon_{p.\lambda} - 2\varepsilon_{a,\lambda} - \varepsilon_b)$

$A_{\lambda,0}$; $A_{\lambda,\infty}$ $A_{\lambda,0} = a_0 \varepsilon_{a,\lambda} d + b_0 \varepsilon_{b,\lambda} d$; $A_{\lambda,\infty} = \frac{a_0}{2} \varepsilon_{p,\lambda} d$

Rate law, general $\frac{dx}{dt} = \frac{1}{2} k_3 (a_0 - 2x)^3$

Rate law, absorbance $\frac{dA_\lambda}{dt} = \frac{4k_3}{q_\lambda^2}(A_{\lambda,\infty} - A_\lambda)^3$

Integrated rate law $$\frac{1}{(A_{\lambda,\infty} - A_\lambda)^2} - \frac{1}{(A_{\lambda,\infty} - A_{\lambda,0})^2} = \frac{8k_3}{q^2} \cdot t \qquad (130)$$

Special cases

1) $A_{\lambda,\infty}=0$; $\varepsilon_{p,\lambda}=0$; $q_\lambda=-(3\varepsilon_{a,\lambda}+\varepsilon_b)d$

2) $A_{\lambda,0}=0$; $\varepsilon_{a,\lambda}=\varepsilon_{b,\lambda}=0$; $q_\lambda=\varepsilon_{p,\lambda}\cdot d$

analogous to reaction type b) 3rd order

Reaction type c) $a+b+c\rightarrow p \quad a_0=b_0=c_0$

Stoichiometric coefficients $\nu_a=\nu_b=\nu_c=-1$; $\nu_p=1$

$q_\lambda=d\sum\nu_i\varepsilon_{i,\lambda}$ $\quad q_\lambda=d(\varepsilon_{p,\lambda}-\varepsilon_{a,\lambda}-\varepsilon_{b,\lambda}-\varepsilon_{c,\lambda})$

$A_{\lambda,0}$; $A_{\lambda,\infty}$ $\quad A_{\lambda,0}=a_0\varepsilon_{a,\lambda}d-a_0\varepsilon_{b,\lambda}d+a_0\varepsilon_{c,\lambda}d$;

$A_{\lambda,\infty}=a_0\varepsilon_{p,\lambda}d$

Rate law, general $\quad \frac{dx}{dt}=k_3(a_0-x)^3$

Rate law, absorbance $\quad \frac{dA_\lambda}{dt}=\frac{k_3}{q_\lambda^2}(A_{\lambda,\infty}-A_\lambda)^3$

Integrated rate law

$$\frac{1}{(A_{\lambda,\infty}-A_\lambda)^2}-\frac{1}{(A_{\lambda,\infty}-A_{\lambda,0})^2}=\frac{2k_3}{q_\lambda^2}\cdot t \qquad (131)$$

$t=0$; $A_\lambda=A_{\lambda,0}$

Special cases

1. $A_{\lambda,\infty}=0$; $\varepsilon_{p,\lambda}=0$;
   $q_\lambda=-(\varepsilon_{a,\lambda}-\varepsilon_{b,\lambda}+\varepsilon_{c,\lambda})d$

2. $A_{\lambda,0}=0$; $\varepsilon_{a,\lambda}=\varepsilon_{b,\lambda}=0$;
   $q_\lambda=\varepsilon_{p,\lambda}\cdot d$

analogous to reaction type b) 3rd order

The reaction type $a+b+c\rightarrow p$ with $a_0\neq b_0\neq c_0$ is very rare in a 3rd order reaction. Therefore, we can forego a further discussion; for more details see [5].

7.1.2.3 Pseudo 1st Order Reactions

In many reactions, particularly bimolecular reactions, there is frequently a considerable excess of one of the reactants. Therefore, the concentration of this substance hardly changes during the reaction. The speed of reaction is determined solely by the concentration of the second component, i.e. the reaction can be regarded as a 1st order reaction and we speak of a "pseudo

1st order" reaction. Examples of this are hydrolysis or solvolysis reactions which are frequently catalyzed by protons and hydroxyl ions. In these cases the rate law is written in the form:

$$\frac{dx}{dt} = k_2(a_0 - x) \cdot b_0 = k_1'(a_0 - x) \tag{132}$$

with

$$k_1' = k_2 \cdot b_0 \ .$$

This result originates from Eqs. (125a, b). However, the rate constant k_1 obtained depends on the concentration b_0, or the pH-value for acid- or base-catalyzed reactions.

An example of such a reaction is discussed in Sect. 7.4.

7.1.2.4 Consecutive Reactions

In the simplest case, these are of the type $a \xrightarrow{k_1} b \xrightarrow{k_2} p$ i.e. each stage is a 1st order reaction and is *irreversible*. We have a complicated reaction if b can revert to a, i.e. if a *reversible* reaction step with the establishment of an equilibrium is involved:

$$a \underset{k_{-1}}{\overset{k_1}{\rightleftharpoons}} b \rightarrow p \ .$$

Such reactions frequently occur in photochemistry, e.g.

trans ⇆ *cis* → cyclization product.

The rapid transition of an excited molecule $^1M^*$ to the triplet state $^3M^*$ with subsequent deactivation to the singlet ground state $^1M^0$ is an analogous reaction:

$$^1M^* \xrightarrow{k_{isc}} {}^3M^* \xrightarrow{k_{PM}} {}^1M^0 \ .$$

All reactions of this type have in common that the change of concentration with time of b (*cis*, $^3M^*$) leads to a rate law in the form of an inhomogeneous differential equation:

$$\frac{db}{dt} = k_1 a - k_2 b \ .$$

Since a is given by

$$a = a_0 e^{-k_1 t}$$

after a 1st order reaction step it follows that:

$$\frac{db}{dt} = a_0 e^{-k_1 t} - k_2 b \ . \tag{133}$$

Fromherz [7] has given an integration of this inhomogeneous differential Eq. (133). Frost and Pearson [3], Mauser [5], Schwetlick et al. [4] and Ebisch et al. [9] have dealt more generally with reversible consecutive reactions.

In more recent work, Derauleau and Dübler [10] have considered the spectrophotometric analysis of a consecutive reaction taking account of an intermediate compound. They have discussed two cases, namely that the rate constants k_1 and k_2 are identical ($k_1 = k_2$) and that they are not identical ($k_1 \neq k_2$). They also discussed the form of the absorbance-time curves for both cases.

.1.2.5 Parallel Reactions

We briefly discuss the simplest type: $a \xrightarrow{k_1} b\ ;\ a \xrightarrow{k_2} c$.

Transformation variables x_1 and x_2 between which the relationship (134) exists are assigned to the two steps [5]:

$$\frac{x_2}{x_1} = \frac{k_2}{k_1} = n\ , \quad x_2 = n \cdot x_1 \ . \tag{134}$$

The rate law follows from a formal description of this reaction:

$$\frac{da}{dt} = -k_1 a - k_2 a = -(k_1 + k_2) a = -k \cdot a \tag{135}$$

with

$$k = k_1 + k_2 \ .$$

The following holds for the concentrations:

$$a = a_0 - x_1 - x_2\ , \quad b = x_1 \quad \text{and} \quad c = x_2 \ .$$

At any time t, taking account of Eq. (134), the absorbance is given by:

$$A_\lambda = a_0 \varepsilon_{a,\lambda} d - (1+n) x_1 \varepsilon_{a,\lambda} d + x_1 \varepsilon_{b,\lambda} d - n x_1 \varepsilon_{c,\lambda} d$$

and

$$A_\lambda - A_{\lambda,0} = q_\lambda \quad \text{with} \quad q_\lambda = d(\varepsilon_{b,\lambda} + n\varepsilon_{c,\lambda} - (1+n)\varepsilon_a) \ ,$$

$$A_{\lambda,\infty} = a_0(\varepsilon_{b,\lambda} - n\varepsilon_{c,\lambda})d \ . \quad \text{results for } t \to \infty \ .$$

If we again assume that:

$$\frac{1}{q_\lambda}\cdot\frac{dA_\lambda}{dt} = \frac{dx_1}{dt} = k(a_0 - x_1) \ ,$$

we obtain with $x_1 = \dfrac{A_\lambda - A_{\lambda,0}}{q_\lambda}$ and expression q_λ

$$\frac{dA_\lambda}{dt} = k(A_{\lambda,\infty} - (1+n)A_{\lambda,t}) \ .$$

Integration provides the solution ($t = 0 \to A_\lambda = A_{\lambda,0}$):

$$\ln\frac{A_{\lambda,\infty} - (1+n)A_\lambda}{A_{\lambda,\infty} - (1+n)A_{\lambda,0}} = -(1+n)kt \ . \tag{136}$$

The values of $A_{\lambda,\infty}$ and $n = k_2/k_1$ are required for the evaluation. Since n gives the concentration ratio of the products directly, it is necessary to determine this ratio spectrophotometrically during or at the end of the reaction, which requires a knowledge of the extinction coefficients $\varepsilon_{b,\lambda}$ and $\varepsilon_{c,\lambda}$. This gives the value of $A_{\lambda,\infty}$ and Eq. (136) can then be evaluated. The situation discussed here represents a time dependent multicomponent system to which we have already referred in Sect. 4.2 [11]. If n is known and a_0 is given then the reaction can be followed by measuring c or b because the relationship $c/b = n$ is valid at any time during the reaction.

In general, the following applies at wavelength λ at any time during the reaction

$$A_\lambda = (a_0 - (1+n)b)\varepsilon_{a,\lambda}d + b\cdot\varepsilon_{b,\lambda}d + nb\cdot\varepsilon_{c,\lambda}d \ .$$

If b and c do not absorb at wavelength λ ($\varepsilon_{b,\lambda} = \varepsilon_{c,\lambda} \neq 0$) it follows that:

$$A_\lambda = a_0\varepsilon_{a,\lambda}d - (1+n)b\cdot\varepsilon_{a,\lambda}d \ ,$$

$$A_{\lambda,0} - A_\lambda = (1+n)b\cdot\varepsilon_{a,\lambda}d$$

and thus

$$b = \frac{A_{\lambda,0} - A_\lambda}{(1+n)\varepsilon_{a,\lambda}d} \quad \text{and} \quad c = n\frac{A_{\lambda,0} - A_\lambda}{(1+n)\varepsilon_{a,\lambda}d} \ .$$

In this case, the reaction can be followed by the decrease in the absorbance of reactant a. However, this requires the determination of n.

7.2 The Number of Linearly Independent Partial Reactions

When setting out the different reactions together with their rate laws it was assumed that we were dealing with uniform, linearly independent reactions. However, in a practical measurement and evaluation of absorbance data (and other physical measurements) obtained as a function of time we must verify whether we are dealing with a uniform reaction with linearly independent sub-reactions. Matrix rank analysis is particularly suited to this purpose. Sternberg, Ainsworth and others have applied it to kinetic measurements [12–16].

When following a reaction by measuring absorbances at several wavelengths (reaction spectra) the measurements $\Delta A_{\lambda,t}$ ($\Delta A_{\lambda,t} = A_{\lambda,t} - A_{\lambda,0}$, see Sect. 7.1.1) can be recorded in the form of a matrix.

We have already made use of this form of representation in Sect. 6.3.2, Eq. (104). In that case, we had to assess a complex-forming equilibrium and wavelength λ and concentration c were the variable parameters. For a kinetic study we employ wavelength λ and time t.

The experimental matrix $\Delta A_{\lambda,t}$ for p different wavelengths and q different times is:

$$\Delta A_{\lambda,t} = \begin{pmatrix} \Delta A_{11} & \Delta A_{12} & \Delta A_{13} & \dots & \Delta A_{1q} \\ \Delta A_{21} & \Delta A_{22} & \Delta A_{23} & \dots & \Delta A_{2q} \\ \dots & \dots & \dots & \dots & \dots \\ \Delta A_{p1} & \Delta A_{p2} & \Delta A_{p3} & \dots & \Delta A_{pq} \end{pmatrix} . \quad (137)$$

If the rank of the matrix can be determined (see e.g. [8, 17]) we can make the following statements about the number of linearly independent reactions:

Rank s = 1: one linearly independent reaction,
Rank s = 2: two linearly independent reactions,
Rank s = 3: three linearly independent reactions.

Examples are:

$s = 1$; *one linearly independent reaction*

$$a \rightarrow b$$
$$a + b \rightarrow c$$
$$a + b \rightarrow c + d$$

$$a \begin{smallmatrix} \nearrow b \\ \searrow c \end{smallmatrix} \quad \text{or} \quad a \rightarrow b + c$$

$s = 2$; *two linearly independent reactions*

$$a \rightarrow b \rightarrow c$$
$$a \rightleftarrows b \rightarrow c$$
$$a + b \rightarrow c \rightarrow d + e$$

$s = 3$; *three linearly independent reactions*

$$a \rightarrow b \rightarrow c \rightarrow d$$
$$a+b \rightleftarrows c+d \rightarrow e \rightarrow f$$

In the case of larger matrices, which are inevitable in these applications, the rank of a matrix can only be determined accurately, with the appropriate computer programs, if the data are free of errors. However, since the experimentally determined pxq-matrix (generally $p \neq q$) is always encumbered with measurement errors, a numerical matrix rank analysis is only useful in practice if further statistical criteria, the advantages and disadvantages of which will be discussed later [18–20], are specified.

These difficulties can be avoided by using the graphical method, known as *graphical matrix rank analysis*, for analyzing kinetic measurements. Mauser [21] has developed this method in connection with a general theory of isosbestic points. Graphical matrix rank analysis was discussed in connection with photometric titration and the investigation of complex-forming equilibria in Sects. 6.2, 6.3.2 and 6.3.3. Therefore, the criteria for the application of matrix rank analysis to kinetic measurements will be discussed only briefly.

a) *ISOSBESTIC POINTS:*

If we record a changing absorption spectrum repetitively over a substantial wavelength region it is possible that these *reaction spectra* intersect precisely at one point at one or several wavelengths. These are the *isosbestic points*, see Figs. 62 and 65. If this is the case, it is very probable that the reaction is uniform, i.e. we are dealing with a *linearly independent reaction* (rank $s = 1$). However, if spectra intersect in a region without passing through an isosbestic point the reaction is not uniform, even if isosbestic points occur in other regions. If there are no crossing points and therefore no isosbestic points at all, we can say *nothing* about the uniformity of a reaction.

b) *ABSORBANCE-TIME DIAGRAM:* $A_{\lambda,t} = f(t)$

Here, the absorbance values measured at wavelength λ are plotted against time. By using the expressions summarized in Sect. 7.1, which were obtained by integrating the appropriate rate laws, it is possible to establish the order and mechanism of a reaction and to determine the rate constant. For a uniform reaction an evaluation at different wavelengths must provide identical results within the limits of error.

c) *ABSORBANCE DIAGRAMS:* $A_i(t) = f(A_1(t))$

If the absorbance values measured at wavelength λ_i are plotted against those measured at wavelength λ_1 straight lines are obtained for a uniform reaction (rank $s = 1$). With ranks ≥ 2 curved plots result.

Mauser and his co-workers have shown that other important information can be obtained from absorbance diagrams [22–24]. This applies par-

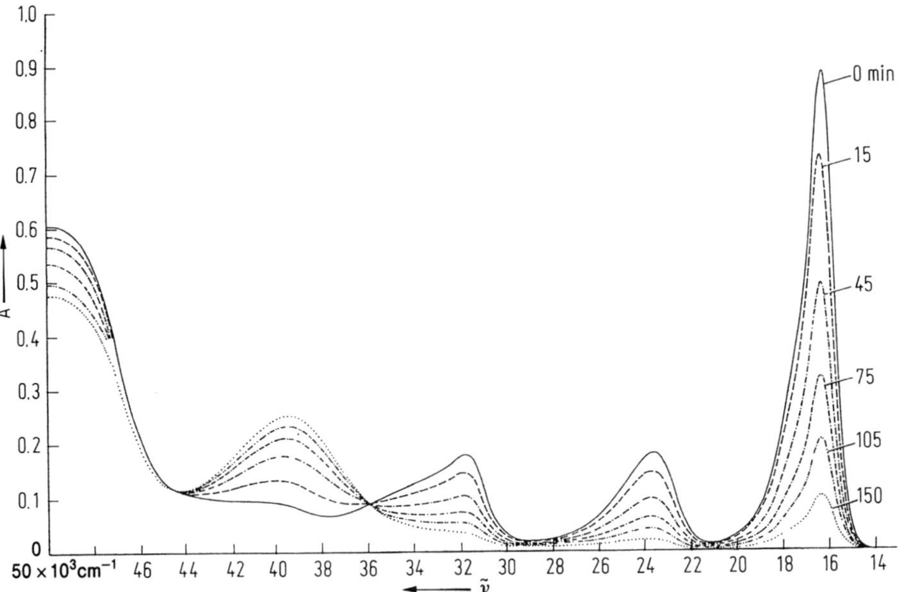

Fig. 62. Reaction spectra of the alkaline hydrolysis of malachite green 50% H_2O/50% buffer pH = 9; $c_0 = 1.657\times10^{-5}$ M; room temperature; Perkin-Elmer model 320, automatic time program; tandem cuvette d = 0.876 cm

ticularly where equilibria and consecutive reactions are involved. If the straight lines of an A-diagram pass through zero we can conclude that only one substance absorbs in the wavelength region used for the measurements.

d) *ABSORBANCE-DIFFERENCE DIAGRAMS:*

$\Delta A_i(t) = f(\Delta A_1(t))$; (ΔA diagram)

The equation for an absorbance-difference diagram has already been presented in Sect. 6.2, Eq. (86). In this form it applies to a uniform reaction, i.e. rank s = 1. In the case of a 2nd order reaction (s = 1) we have:

$$A_1 = a_0\varepsilon_{a1}\cdot d + b_0\varepsilon_{b1}\cdot d + (\varepsilon_{c1} + \varepsilon_{d1} - \varepsilon_{a1} - \varepsilon_{b1})\cdot d\cdot x = A_1^0 + q_1\cdot x \ ,$$

$$\Delta A_1 = q_1 x \ .$$

$$A_i = a_0\varepsilon_{ai}\cdot d + b_0\varepsilon_{bi}\cdot d + (\varepsilon_{ci} + \varepsilon_{di} - \varepsilon_{ai} - \varepsilon_{bi})\cdot d\cdot x = A_i^0 + q_i x \ ,$$

$$\Delta A_i = q_i x \ .$$

Eliminating x in both equations gives

$$\Delta A_i = \Delta A_1\cdot\frac{q_i}{q_1} = z_i\cdot\Delta A_1 \ . \qquad (138)$$

If the relevant values of ΔA_i are plotted against ΔA_1 (λ_1 = reference wavelength) for different wavelengths λ_i (i = 2, 3...) we obtain straight lines passing through the origin, the slopes of which are given by the ratio q_i/q_λ. This also presents an easy way of testing the uniformity of a reaction. However, the method requires the plotting of as many wavelength combinations as possible over a long spectral range extending into the further UV region. In the case of two and more linearly independent partial reactions, if only one component absorbs in the spectral region used for an evaluation, it will always simulate a uniform reaction with rank s = 1 [25]. This is illustrated by the hydrolysis of *p*-nitrophenyl acetate [26].

If the extinction coefficients are known, the ratio q_i/q_1 corresponding to the definition of the q_λ values can be calculated. This ratio can then be compared with the slope obtained from the diagram [27]. This also verifies that the reaction has proceeded as assumed.

Curved absorbance-difference diagrams or absorbance diagrams frequently occur if $s \geqslant 2$. In some cases for s = 2, they show two linear components [27–29] which indicates that the sub-reactions are 1st order reactions.

e) *ABSORBANCE-DIFFERENCE-QUOTIENT DIAGRAMS* (ΔAQ):

Consider a system where we have precisely two linearly independent partial reactions. Then the following applies for the absorbance at wavelength "1".

$$A_1 = [(a_0 - x_1 - x_2)\varepsilon_{a,1} + x_1\varepsilon_{b,1} + x_2\varepsilon_c]\,d \ .$$

Rearranging to show the coefficients of x_1 and x_2 gives

$$\Delta A_1 = A_1 - A_{01} = d(\varepsilon_{b,1} - \varepsilon_{a,1})x_1 + d(\varepsilon_{c,1} - \varepsilon_{a,1})x_2 = q_{11}x_1 + q_{12}x_2$$

and analogously for wavelengths 2 and 3:

$$\Delta A_2 = q_{21}x_1 + q_{22}x_2 \ ,$$

$$\Delta A_3 = q_{31}x_1 + q_{32}x_2 \ .$$

As shown in Sect. 6.3, after solving the first two equations for x_1 and x_2 and substituting the solutions into the third equation we obtain the expression:

$$\Delta A_3 = \frac{D_{23}}{D_{12}}\Delta A_1 + \frac{D_{13}}{D_{12}}\cdot\Delta A_2 \ .$$

If ΔA_1 is brought to the left-hand side we have:

$$\frac{\Delta A_3}{\Delta A_1} = n + m\cdot\frac{\Delta A_2}{\Delta A_1} \qquad (139)$$

with

$$n = \frac{D_{23}}{D_{12}} \quad \text{and} \quad m = \frac{D_{13}}{D_{12}} \; .$$

$$D_{12} = q_{11} q_{22} - q_{12} q_{21} \; ; \quad D_{23} = q_{31} q_{22} - q_{32} q_{21} \; ,$$

$$D_{13} = q_{11} q_{32} - q_{31} q_{12} \; .$$

If $\Delta A_3/\Delta A_1$ is plotted against $\Delta A_2/\Delta A_1$ a straight line results in the case of two linearly independent sub-reactions. Naturally, this can also be used to determine the number of sub-reactions.

It can only be decided with certainty whether $s = 2$ if as many wavelength combinations as possible are used for the evaluation. They should cover a broad spectral range and preferably extend into the UV region. If a straight line passing through the origin results or lines parallel to one of the axes are found, then the ΔAQ diagram is degenerate and we can make no statement about the number of linearly independent sub-reactions.

This method can also be extended to include the case of three independent sub-reactions [30].

Gauglitz has given a clear description of the methods for evaluating spectrophotometric measurements of reactions [30, 31]. This general work also illustrates the limitations of the different methods.

7.3 Evaluation of Kinetic Measurements

The rate laws and their solutions discussed in Sect. 7.1 permit, in principle, an evaluation of kinetic measurements. The advantage of spectrophotometric methods over other methods in the measurement of kinetics arises because, by means of reaction spectra, a great number of wavelengths can be used for the evaluation of one and the same experiment. This leads to improved results.

Furthermore, a reaction spectrum also offers the opportunity of selecting the most suitable wavelength region for an evaluation, which is particularly advantageous if only the reactant or product absorbs at a specific wavelength. Each of these cases has been referred to.

However, an exact evaluation using the above equations frequently requires a knowledge of the values of $A_{\lambda,\infty}$ and $A_{\lambda,0}$. The absorbance at $t = 0$ can usually be measured, e.g. by use of a tandem cuvette, or calculated by a suitable procedure from the initial concentrations and extinction coefficients. However, this is much more difficult, and often not possible, for $A_{\lambda,\infty}$ values.

It is frequently the case that the initial values of a reaction cannot be measured accurately enough. Therefore, evaluation using an integrated rate law can result in very divergent values in the initial region.

For this reason, many methods which circumvent these difficulties have been described in the literature. Mauser has given a summary, Ref. [5] chapter III. Because of their importance, methods of evaluation based on difference equations should be mentioned briefly [32].

In the general case of a 1st order reaction, Eq. (125a) applies

$$A_{\lambda,\infty} - A_{\lambda,t} = (A_{\lambda,\infty} - A_{\lambda,0})\, e^{-k_1 t} \; . \tag{a}$$

At time $t+\Delta t$

$$A_{\lambda,\infty} - A_{\lambda,t+\Delta t} = (A_{\lambda,\infty} - A_{\lambda,0})\, e^{-k_1(t+\Delta t)} \; . \tag{b}$$

Subtraction of equation (b) from (a) gives:

$$A_{\lambda,t+\Delta t} - A_{\lambda.t} = (A_{\lambda.\infty} - A_{\lambda.0} e^{-k_1 t}(1 - e^{-k_1 \Delta t}) \; . \tag{c}$$

Since

$$(A_{\lambda,\infty} - A_{\lambda,0})\, e^{-k_1 t} = A_{\lambda,\infty} - A_{\lambda,t}$$

we finally obtain the equation:

$$A_{\lambda,t+\Delta t} = A_{\lambda,\infty}(1 - e^{-k_1 \Delta t}) + A_{\lambda,t}\, e^{-k_1 \Delta t} \; . \tag{140}$$

If Δt is kept constant and $A_{\lambda,t+\Delta t}$ plotted against $A_{\lambda,t}$ a straight line is obtained. Equation (140) corresponds to the equation given by Swinbourne for evaluating a 1st order reaction [33].

Equation (140) can be solved without knowledge of the initial values and the value of $A_{\lambda,\infty}$.

From Eq. (140), the latter value is obtained at $t \to \infty$ as $A_{\lambda,t+\Delta t} = A_{\lambda,t} = A_{\lambda,\infty}$. Since the slope yields $e^{-k_1 \Delta t}$, k_1 can be determined directly.

Starting with equation (c) above we can also write:

$$A_{\lambda,t+\Delta t} - A_{\lambda,t} = (A_{\lambda,0} - A_{\lambda,\infty})(e^{-k_1 \Delta t} - 1)\, e^{-k_1 t} \; .$$

The logarithm of this equation is the Guggenheim relationship for evaluating a 1st order reaction [34], i.e.:

$$\ln(A_{\lambda,t+\Delta t} - A_{\lambda,t}) = -k_1 t + \ln\,[(A_{\lambda,0} - A_{\lambda,\infty})(e^{-k_1 \Delta t} - 1)] \; . \tag{141}$$

Equations (140) to (141) show that the solution is independent of the $A_{\lambda,0}$ and $A_{\lambda,\infty}$ values, i.e. the uncertainty of the data at the beginning and end of the reaction has been eliminated.

For 2nd order reactions, appropriate expressions which allow an evaluation independent of $A_{\lambda,\infty}$ have already been given for the cases

$2a \to p$ and $a+b \to p$ $(a_0 = b_0)$

see Eqs. (126a, 127b). Mauser has recently discussed in detail the evaluation of spectrophotometric measurements for 2nd order reactions without additional information, i.e. without knowledge of the extinction coefficients [35].

It is also possible to evaluate kinetic measurements by means of the method of *formal integration* [36]. From Eq. (125), the following applies to a 1st order reaction:

$$dA_{\lambda,t} = k_1 A_{\lambda,\infty} dt - k_1 A_{\lambda,t} dt \ .$$

If this equation is integrated formally between the limits t and t′ we obtain:

$$A_{\lambda,t'} - A_{\lambda,t} = k_1 A_{\lambda,\infty}(t'-t) - k_1 \int_t^{t'} A_{\lambda,t} dt$$

or

$$\frac{A_{\lambda,t'} - A_{\lambda,t}}{t'-t} = k_1 A_{\lambda,\infty} - k_1 \frac{\int_t^{t'} A_{\lambda,t} dt}{t'-t} \ . \tag{142}$$

If the left-hand side of Eq. (142) is plotted against the integral divided by the time difference $t'-t$ a straight line is obtained whose slope provides k_1 directly and its ordinate intercept, $A_{\lambda,\infty}$, see the example in Sect. 7.4.

This method of formal integration has the advantage that it is almost unaffected by errors of measurement, whilst an indirect evaluation of the differential equations and their solutions is made more difficult, i.e. far less certain, by such errors.

In contrast to the other graphical evaluation methods discussed above, this method has a further advantage because the integration limits t′ and t can be selected at random and are not required to correspond to a constant difference Δt. Lachmann et al. have given examples of the application of this method to 1st order reactions [26].

For a 2nd order reaction with equal initial concentrations $a_0 = b_0$ (see Eq. 127), the formal integration provides

$$A_{\lambda,t'} - A_{\lambda,t} = \frac{k_2}{q_\lambda} A^2_{\lambda,\infty}(t'-t) - 2A_{\lambda,\infty} \frac{k_2}{q_\lambda} \int_t^{t'} A_{\lambda,t} dt + \frac{k_2}{q_\lambda} \int_t^{t'} A^2_{\lambda,t} dt$$

or

$$\frac{A_{\lambda,t'} - A_{\lambda,t}}{t'-t} = Z_1 - Z_2 \frac{\int_t^{t'} A^2_{\lambda,t} dt}{t'-t} + Z_3 \frac{\int_t^{t'} A^2_{\lambda,t} dt}{t'-t} \ , \tag{143}$$

$$Z_1 = \frac{k_2}{q_\lambda} A^2_{\lambda,\infty} \ , \qquad Z_2 = 2A_{\lambda,\infty} \cdot \frac{k_2}{q_\lambda} \ , \qquad Z_3 = \frac{k_2}{q_\lambda} \ .$$

From this the equations for $A_{\lambda,\infty}$ and k_2 are obtained immediately

$$\frac{Z_1}{Z_3} = A_{\lambda,\infty}^2 \quad \text{or} \quad \frac{Z_2}{2Z_3} = A_{\lambda,\infty} \ . \tag{144}$$

Starting with the differential Eq. (135) ($a_0 \neq b_0$) also leads to Eq. (143) but the coefficients Z_1, Z_2 and Z_3 are defined as follows in this case:

$$Z_1 = \frac{k_2}{q_\lambda} A_{\lambda,\infty} \cdot A'_{\lambda,\infty} \ , \quad Z_2 = \frac{k_2}{q_\lambda}(A_{\lambda,\infty} + A'_{\lambda,\infty}) \ , \quad Z_3 = \frac{k_2}{q_\lambda} \ . \tag{144a}$$

From this result expressions for $A_{\lambda,\infty}$ and $A'_{\lambda,\infty}$ can also be found. However, the values calculated are frequently very inaccurate.

These methods of evaluation, and those discussed in Sect. 7.2 for graphical matrix rank analysis, require high photometric accuracy if definite conclusions about the uniformity of a reaction are to be drawn or reliable values for the rate constants obtained. For this reason, Lachmann has proposed a universal reaction system which has good repeatability and is suitable for testing a UV-VIS spectrophotometer [37]. It involves the hydrolysis of 2-hydroxy-5-nitro-α-toluene sulfonic acid sultone:

$$O_2N\text{-}C_6H_3(CH_2SO_2O) \xrightarrow{H_2O} O_2N\text{-}C_6H_3(CH_2SO_3^{\ominus})(\bar{\underline{O}}|^{\ominus}) + 2\,H^{\oplus}$$

The rate of hydrolysis depends strongly upon the pH-value, buffer composition and ionic strength. Consequently, the half-life of the reaction can be varied from milliseconds to several hours by the appropriate selection of these parameters. Thus, this reaction is suitable for testing conventional UV-VIS as well as stopped-flow spectrophotometers [38].

Kinetic investigations by spectrophotometric methods require that absorbance values be measured at timed intervals. By following a reaction at one wavelength alone the data can be surveyed at a glance; they are available for further analysis if a table is drawn up. However, these methods of evaluation have shown that a single-wavelength study is frequently inadequate since we cannot determine the uniformity of a reaction in that way. For that reason, it is appropriate to measure the absorption spectra at timed intervals over a spectral region which is as broad as possible. The spectra obtained in this way are known as reaction spectra and, by the presence of isosbestic points, they may indicate that we are probably dealing with a uniform reaction. A measurement of such spectra with modern recording UV-VIS spectrophotometers causes no problems since these instruments usually have microprocessor control. Thus, the repeat times appropriate to the velocity of a reaction can be specified. As far as the evaluation is concerned, this means that numerous data, which could scarcely be evaluated manually in an acceptable time can be rapidly produced. Therefore, the linking of a microcomputer to the spectrophotometer is recommended for further processing of the data. Most spectrophotometers are fitted with an appropriate interface for this purpose. Several suggestions have been made for linking computers to older equipment [39–41].

7.4 Examples

A large proportion of all solvolysis or hydrolysis reactions are pseudo 1st order reactions. The base-catalyzed reaction of malachite green is our first example. It changes to its colorless carbinol base in aqueous solution at pH 9:

$(CH_3)_2\overset{\oplus}{N}$ C $N(CH_3)_2$ $\xrightarrow[H_2O]{OH^{\ominus}}$ $(CH_3)_2N$ COH $N(CH_3)_2$

This reaction can be followed conveniently spectrophotometrically within 4 h at room temperature. Figure 62 shows the reaction spectra which indicate a clearly defined isosbestic point at 36000 cm^{-1} = 277.8 nm. This suggests a uniform reaction. From the reaction spectra we can see that 4 absorption maxima and one shoulder on the short wavelength side of the intense long-wavelength absorption band are available for analysis, i.e., $\lambda_1 = 615$ nm (16260 cm^{-1}); λ_2 (sh) = 575 nm (17400 cm^{-1}); $\lambda_3 = 425$ nm (23530 cm^{-1}); $\lambda_4 = 315$ nm (31750 cm^{-1}) and $\lambda_5 = 255$ nm (39215 cm^{-1}).

The absorbance decreases continuously with time at wavelengths λ_1 to λ_4 but it increases continuously at λ_5. Furthermore, it does not equal zero when t = 0.

A tandem cuvette was used for following the reaction. At t = 0 both solutions (malachite green in H_2O and the solution of pH 9) were in the separate chambers. After setting the temperature (T = 293 K) both solutions were mixed rapidly outside the instrument, returned to the temperature-controlled cuvette holder of the spectrophotometer and the recording of the reaction spectra started immediately. The spectra were recorded on the chart and also stored on a disk using an interfaced Apple II computer. Subsequently, they were printed out as $A_{\lambda,t} = f(\lambda)$.

Testing of the uniformity of the reaction by means of an absorbance diagram and absorbance-difference diagram confirms that the reaction is uniform and of rank s = 1.

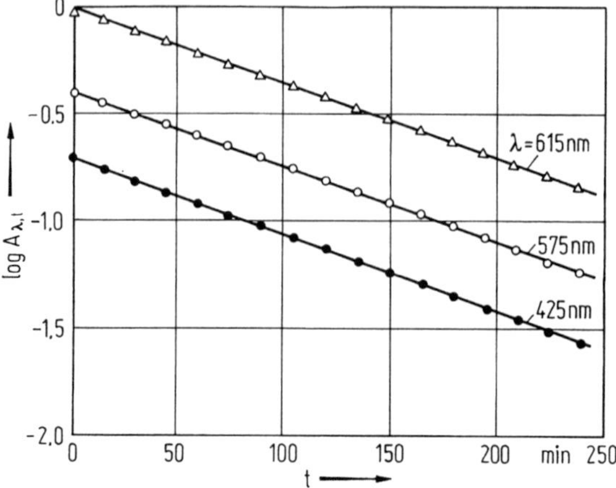

Fig. 63. Kinetic evaluation of the malachite-green hydrolysis using Eq. (127a)

Table 19. k_1-Values for the malachite-green reaction, water pH = 9; T = 293 K

Evaluation using:	λ	k_1
Eq. (125a)	615 nm	$1.36 \times 10^{-4}\,s^{-1}$
	575 nm	$1.34 \times 10^{-4}\,s^{-1}$
	425 nm	$1.38 \times 10^{-4}\,s^{-1}$
Eq. (125) $A_{\lambda,\infty} = 0.300$	255 nm	$1.34 \times 10^{-4}\,s^{-1}$
Swinbourne Eq. (140)	615 nm	$1.34 \times 10^{-4}\,s^{-1}$
Formal integration Eq. (142)	615 nm	$1.24 \times 10^{-4}\,s^{-1}$

In order to determine the rate constant k_1 from the data, log $A_{\lambda,t}$ was plotted against t according to Eq. (125a). Figure 63 shows that parallel straight lines are obtained for λ_1. λ_2 and λ_3 and k_1-values can be calculated from their gradients. This simple procedure is not possible for λ_5. For this wavelength Eq. (125) was used and the required $A_{255,\infty}$-value was estimated, by extrapolation of the data for $t \to \infty$, as $A_{\lambda,\infty} = 0.3$. Using this value we also obtain a straight line for ln $(A_{\lambda,\infty} - A_{\lambda,t})$ plotted against t. Table 19 shows the good agreement of the k_1-values. This also applies to an evaluation of the data by the Swinbourne difference method, Eq. (140). Figure 64 illustrates an analysis by means of the formal integration method, Eq. (142). The integral $\int_t^{\nu} A_{\lambda,t}\,dt$ was determined by summing with constant steps of $\Delta t = 15$ min and also with increasing Δt. The integrals were obtained using the trapezium rule [42]. Figure 64 shows that the first method (Δt = const = 15 min) gives values which have a systematic error near the end of the reaction. This arises because the absorbance values are very small and the error when calculating the area under the curve by means of the trapezium rule is relatively large.

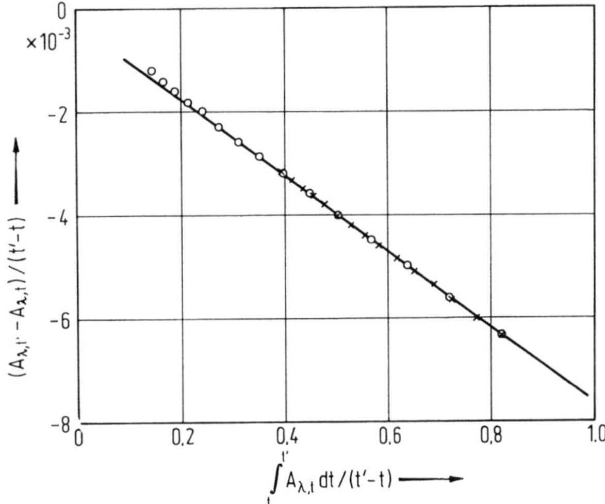

Fig. 64. Evaluation of the malachite-green hydrolysis by means of formal integration, Eq. (142)

In contrast, evaluation by the second method provides considerably more reliable values as can be seen in Fig. 64 (experimental data as crosses).

The base-catalyzed lactone ring opening of coumarin-3-carboxylic acid can be briefly discussed as a further example. This reaction depends on the solvent, the OH-ion concentration and the ionic strength of a solution [43] and follows the reaction scheme:

$COO^{\ominus}$ $\xrightarrow[CH_3OH/H_2O]{OH^{\ominus}}$ $COO^{\ominus}$, $COO^{\ominus}$, $O^{\ominus}$

$COO^{\ominus}$ $+ CO_2$, $O^{\ominus}$

The reaction spectra at T = 293 K (Fig. 65) show two isosbestic points at $\lambda_1 = 333$ nm (30030 cm^{-1}) and at $\lambda_2 = 277$ nm (36100 cm^{-1}) and we can draw the conclusion that we are dealing with a uniform reaction.

Consequently, the decarboxylation of the dicarboxylic acid as a consecutive reaction at 293 K can be excluded. Furthermore, the ΔA-diagrams (Fig. 66) show straight lines passing through the origin for all wavelengths, as is expected for a uniform reaction of rank s = 1. Evaluation of the data was carried out using the Swinbourne equation, Eq. (140); the Guggenheim

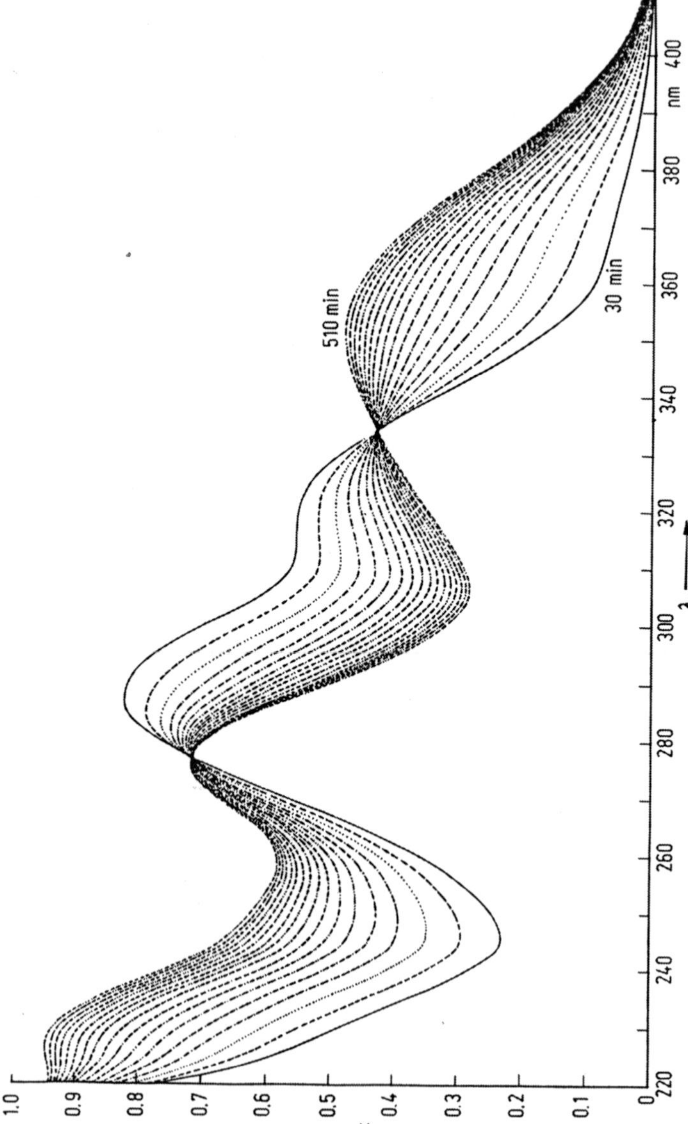

Fig. 65. Reaction spectra of the alkaline lactone-ring opening of coumarin-3-carboxylic acid; c = 1.638×10^{-4} M; solvent: 50% methanol/50% 10^{-2} M NaOH; tandem cuvette d = 0.876 cm; room temperature; Perkin-Elmer model 320; automatic recording at 30 min intervals

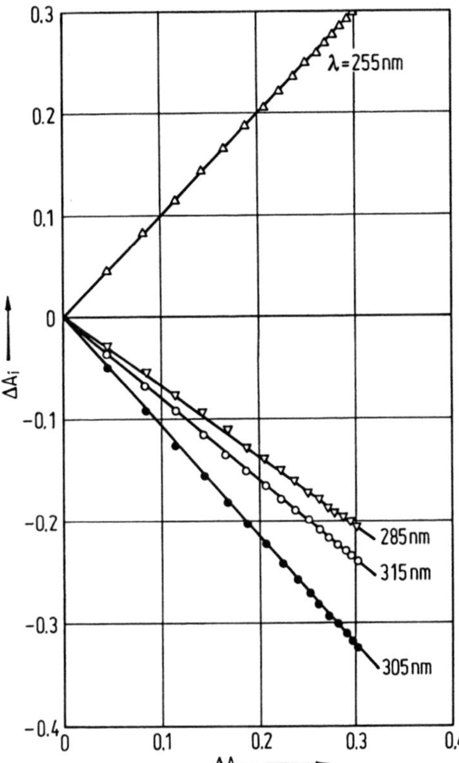

Fig. 66. ΔA-diagram of the reaction spectra in Fig. 65

Table 20. k_1' values for the pseudo 1st order lactone-ring opening of coumarin-3-carboxylic acid; T = 293 K; 50% CH_3OH + 50% NaOH (10^{-2} M)

$k_1' \times 10^5$ [s^{-1}] using: λ |nm|	Eq. (140)	Eq. (141)	Eq. (142)
350	7.36	7.42	7.61
315	7.48	7.52	7.54
305	7.45	7.49	7.53
285	7.11	7.20	7.21
255	7.36	7.37	7.63
k_1'	7.35	7.45	7.50

equation, Eq. (141); and the formal integration method, Eq. (142) at five different wavelengths. The results are summarized in Table 20.

The example of the lactone-ring opening discussed here is representative of the numerous coumarin derivatives for which, on account of their importance, the kinetics have been described in detail [43–46]. Accurate spectroscopic investigation shows that most reactions can be described as a pseudo 1st order with respect to the coumarin derivative.

However, the rate constant k_1 depends on the OH-ion concentration and the overall reaction is 2nd order. Thus, the following applies:

$$k_2[\text{coum}]\times[\text{OH}^-] = k_1' \, [\text{coum}]$$

or

$$k_2 = k_1' \cdot [\text{OH}^-]^{-1} \; .$$

Introducing the ionic product of water, K_w, and the H^+-ion concentration we obtain

$$k_2 = k_1' \, \frac{[\text{H}^+]}{K_w}$$

or in logarithmic form:

$$\log k_2 = \log k_1' + pK_w(T) - pH(T) \; . \tag{145}$$

As shown in Eq. (145), k_2 can be calculated from k_1' at a given temperature. If the k_2-values are required from the k_1'-values determined at different temperatures, the temperature dependence of the pH-values, which is generally not accurately known, causes uncertainty in the k_2-values. However, the temperature dependence of the pK_w-value is well established [47]. Thus, it is worthwhile to determine the k_2-values directly by another method. This requires that the concentrations of the reactants must be comparable and this is frequently too high for spectrophotometric measurement. Conductivity measurements are suitable for following a 2nd order reaction of the coumarin derivatives which have been briefly discussed here [44].

As an example of a 2nd order reaction, the formation of a Schiff's base from *p*-hydroxy aniline (I) and pyridine-4-aldehyde (II) can be briefly discussed:

$$\underset{\text{I}}{\text{HO}-C_6H_4-\text{NH}_2} + \underset{\text{II}}{\text{O}=\overset{\text{H}}{\overset{|}{\text{C}}}-C_5H_4\text{N}} \rightarrow \underset{\text{III}}{\text{HO}-C_6H_4-\text{N}=\text{CH}-C_5H_4\text{N}} + \text{H}_2\text{O}$$

Figure 67 shows the reaction spectra at T = 293 K which were used for following the reaction up to 340 min. We calculate $A_{350,\infty} = 4.138$ by means of the known spectroscopic data for the product (III) $\varepsilon_{max} = 11\,930 \text{ l mol}^{-1}\text{ cm}^{-1}$ at $\lambda_{max} = 350$ nm and the given initial concentration $a_0(c_I) = 3.96\times10^{-4}$ M. By comparison with the measured extinction at 350 nm after 340 min ($A_{\lambda,t} = 0.8344$) this corresponds to a reaction of ca. 20%. The spectra show no isosbestic point. However, the ΔA diagram

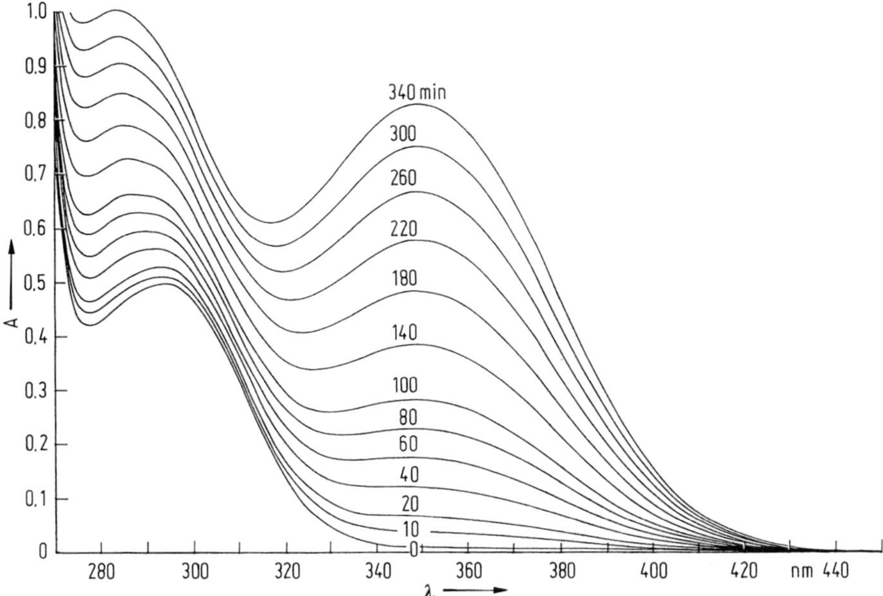

Fig. 67. Reaction spectra for the formation of a Schiff's base from *p*-hydroxyaniline (I) and pyridine-4-aldehyde (II); T = 293 K; solvent: methanol; $c_I = 3.96\times10^{-4}$ M; $C_{II} = 3.96\times10^{-3}$ M; tandem cuvette d = 0.876 cm; Perkin-Elmer model 320; recorded at 10, 20 and 40 min intervals

provides straight lines passing through zero for five wavelengths in the region of both absorption bands. Thus, the conclusion can be drawn that the reaction is uniform.

Only the product (Schiff's base III) absorbs in the long wavelength region $\lambda_{max} = 350$ nm. Thus, the measurements can be evaluated using Eq. (128) in the form:

$$\lg \frac{A'_{c,\infty} - A_{c,t}}{A_{c,\infty} - A_{c,t}} = (b_0 - a_0)\,k_2 t + \lg \frac{b_0}{a_0} \ .$$

The result is illustrated in Fig. 68. A rate constant of $k_2 = 2.83\times10^{-3}\,s^{-1}\,mol^{-1}$ was obtained from the slope of the straight line. When t = 0, the ordinate intercept is at log (b_0/a_0). With $a_0 = 3.96\times10^{-4}$ M and $b_0 = 3.96\times10^{-3}$ we obtain $\log(b_0/a_0) = \log 10 = 1$. Figure 68 shows that the straight line intersects the ordinate at $\log(b_0/a_0) = 1$.

The ratio of concentrations was such that there was a tenfold excess of pyridine-4-aldehyde over *p*-hydroxyaniline. Thus, the data can be evaluated formally as pseudo 1st order. As shown in Eq. (125b), the value of $A_{\lambda,\infty}$ is again required; a value of $A_{\lambda,\infty} = 4.138$ ($\lambda = 350$ nm) (see above) was found.

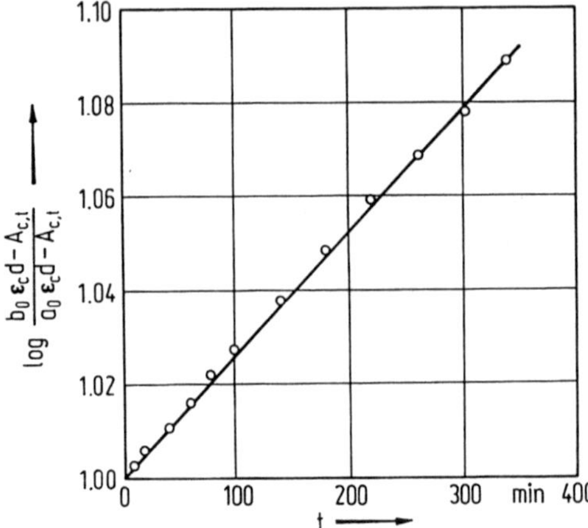

Fig. 68. Kinetic evaluation of the reaction spectra shown in Fig. 67 using Eq. (128); $\lambda = 350$ nm

By plotting $\ln(A_{350,\infty} - A_{350,t})$ against time t a straight line is again obtained with slope $k_1' = k_2 b_0$. From this $k_2 = 2.76 \times 10^{-3}\,s^{-1}\,l\,mol^{-1}$, in good agreement with the evaluation using Eq. (128).

7.5 Fast Reactions

7.5.1 Flow Methods: The Stopped-Flow Technique

In all kinetic experiments the mixing of the various initial components always involves a certain dead-time which elapses before the first kinetic measurements can be reliably determined. In the reactions previously discussed, both solutions were mixed in one cuvette with a dead-time of ca. 5 to 10 s. The dead-time can be reduced below this value to times in the region of 1 to 3 ms by using specific mixing equipment which is based on *flow methods.* This provides the link to the time scale of slow reactions since reactions which lie in the range $10^{-3} < t < 60$ s can be measured by means of flow methods. Thus, a total range of $10^{-3} - 10^5$ is accessible. However, the upper value is essentially limited by the long-term stability of the measuring apparatus used.

There are basically two different flow techniques. The reaction of a mixture is observed either at a constant flow rate at a specific position in the flowing stream or the flow is interrupted suddenly. The latter is the *stopped-flow technique* in which the further progress of the reaction in the volume under observation can be monitored. In both cases, the reactants must be mixed rapidly and effectively in specially designed mixing chambers.

A flow-through cuvette is an open reaction system through which a reaction mixture of known initial concentration is passed at constant velocity.

The composition of the mixture is determined analytically at a specific position, the reaction or observation zone, in the flowing stream.

The basic equation for a continuous flow system may be written in the following notation. We denote the flow rate by u (volume per time unit), the volume element within which the reaction continues by $dV = Q.\ dl$ (Q = cylindrical cross-section) and the concentration of reactants i by c_i at the point of entry into dV and by $c_i + dc_i$ at the exit from dV. Further, we note that, within the volume element dV, the concentration of component i changes due to the reaction. We denote the reaction rate by $\dot{c}_i (\equiv dc_i/dt)$, and so obtain the change with time of the molarity of component i in the volume element as:

$$\frac{dn_i}{dt} = \dot{c}_i \, dV - u \, dc_i \ . \qquad (146)$$

The concentrations of all reactants in the volume element generally reach a constant final value after a specific time. There is then a stationary state in the reaction zone under observation, i.e. $dn_i/dt = 0$ and the following applies:

$$\dot{c}_i \, dV = u \, dc_i \ .$$

The integration of this equation gives:

$$\frac{V}{u} = \int_{c_0}^{c} \frac{dc_i}{\dot{c}_i} \ .$$

If, for example, we take the rate law which applies to 1st order reactions $\dot{c}_i = -k_1 c_i$, we obtain after integration:

$$k_1 = \frac{u}{V} \cdot \ln \frac{c_0}{c} \ . \qquad (147)$$

Since u is expressed as $cm^3\, s^{-1}$ and V as cm^{-3}, the ratio u/V has the dimensions of reciprocal time and Eq. (147) corresponds to the rate law of a 1st order reaction:

$$k_1 = \frac{1}{t} \cdot \ln \frac{c_0}{c} \ .$$

In general, it can be shown that, for reactions of any order, the integrated rate law for a closed system also applies to reactions in a flowing stream, provided that there is no back-mixing. However, instead of the time t, the quotient u/V is the variable. Since $V = Ql$ we can also say that instead of the time as the variable, the path coordinate, l, is the variable. This shows clearly that, in order to follow the rate of reaction in a flowing system, the

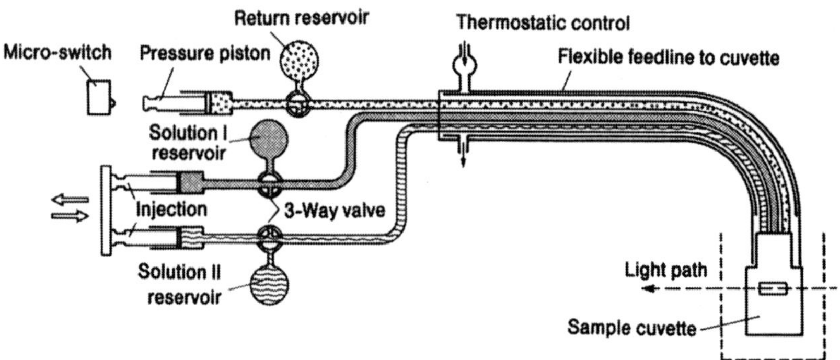

Fig. 69. Stopped-flow accessory: HI-TECH model SFA-11 [49]

reaction must be measured at several points at specific distances, l, from the mixing chamber (corresponding to the timed interval t) at constant flow velocity, u. Thus, only one point at a time can be measured when using this simple arrangement which means that the measurements are extremely time-consuming.

The flow technique outlined here was introduced to spectrophotometry in 1923 [45–47].

Instead of moving the zone under observation, another version of this technique uses a continuous variation of the flow rate and the course of a reaction can be followed as a concentration-time curve [48].

In recent years, the *stopped-flow technique* has increasingly gained in importance when compared with the flowing stream method. In this method, separate solutions of the reactants are mixed very rapidly (within a few ms) in a mixing chamber and the flow is suddenly stopped. Figure 69 shows schematically a simple thermostated version which can be easily fitted as an option to any spectrophotometer [49].

The mixing chamber is a cuvette with a very small volume which permits both absorption and fluorescence measurements. In contrast to a flowing stream, the stopped-flow technique has the advantage that the further progress of a reaction with time can be followed. Thus, a total concentration-time curve or absorbance-time curve can be obtained in one experiment. Depending on the speed of the reaction, an oscilloscope or a fast recorder can be used to record the measurements. Chance [50] and Gibson [51] introduced this technique. Caldin [52], Hague [53], Roughton and Chance [54] and Chance [55] have given more detailed descriptions. Caldin et al. have described a special stopped-flow spectrometer [56].

Apart from the sultone hydrolysis [37] referred to previously, Below et al. have described the reaction of $FeCl_3$ with KSCN [57] in detail.

The use of stopped-flow techniques is particularly important if a complete reaction has to be subdivided into several main steps during its course. This occurs frequently in enzyme-catalyzed reactions as was shown by

Lachmann in the case of the hydrolysis of *p*-cinnamic acid *p*-nitrophenylester catalyzed by α-chymotrypsin [58, 59].

The stopped-flow technique and flow-through cuvette measurements have the disadvantage that in both cases measurements can be made at only one wavelength and that frequently single-beam instruments must be used. For that reason, double-beam rapid spectrometers, which enable the spectrum of the reaction mixture to be measured, have been described in the literature. These instruments have fast moving optical elements which make possible the measurement of a reaction spectrum within 1 to 10 ms [60–63].

The stopped-flow Vidicon spectrometer is another version. The stopped-flow mixing cuvette of this instrument is combined with a polychromator and a two-dimensional diode array (Vidicon) [61, 64, 65]. A corresponding Vidicon spectrometer for routine analysis was described in Chapter 3, see also Sect. 7.7. A Vidicon spectrophotometer has the disadvantage that the cuvette is fitted in front of the entrance slit so that the undispersed light falls with its total intensity onto the sample. This can cause considerable interference in photochemically unstable systems. The stopped-flow HI-TECH spectrometer model SF-4 has been developed for measurements at variable temperatures in the range of $173 < T < 373$ K [65].

7.5.2 Spectroscopic Relaxation Techniques

In following reactions which are faster than those described above, the necessary mixing of the reactants is the limiting time factor. Therefore, another principle must be used when investigating faster reactions ($t < 10^{-3}$ s). In conventional kinetic measurements, we follow the transition from the initial state ($t = 0$) to the final state ($t = \infty$) with respect to time. In contrast, in *relaxation techniques*, the final state, which is generally a condition of equilibrium, is used as the initial state.

If this state is *temporarily* disturbed, the system tends to move to a new condition of equilibrium, i.e. to undergo a chemical relaxation process. A number of spectroscopic relaxation techniques have been based on this principle. Since equilibrium is a function of the variables temperature and pressure, the *temperature-jump* [66] and *pressure-jump techniques* [67] are of particular importance. The former is the more common method for work with solutions. Eigen and De Maeyer have given a detailed description of these techniques [68].

Ertl and Gerischer have described the principles of the temperature-jump technique [69].

The general application of the temperature-jump technique is based on the fact that reaction enthalpy ΔH_0 is finite during chemical changes, i.e. $\Delta H_0 \neq 0$. Thus, a temperature increase generally results in a displacement of equilibrium and hence in a change of the chemical composition of a system. The van t'Hoff equation gives the relative change of the equilibrium constant $\Delta K/K$. In the case of a small temperature change, ΔT:

$$\frac{\Delta K}{K} = \frac{\Delta H_0}{RT} \cdot \frac{\Delta T}{T} \quad . \tag{148}$$

Thus, the relative change of the equilibrium constant is proportional to the relative temperature change. This causes a continuous change of the chemical composition until the system has settled down to the new condition of equilibrium at temperature $T+\Delta T$. Therefore, the effect is a function of the stoichiometry and the original concentrations of the reactants and products [66].

If the profile of the temperature jump is assumed to be rectangular, then equilibrium is re-established according to the simple exponential function:

$$\frac{dc_t}{dt} = \frac{c_0 - c_t}{\tau} \; , \qquad c_t = c_\infty (1 - e^{-t/\tau}) \tag{149}$$

where c_t is the deviation from the equilibrium, c_0 is the sudden displacement of the equilibrium when $t = 0$ and τ is the relaxation time. But the relaxation time is related to the rate constants k_{ij} of the reaction steps which proceed in the direction of attainment of the new equilibrium.

For the simple system

$$A + B \underset{k_{21}}{\overset{k_{12}}{\rightleftharpoons}} AB$$

there results

$$\tau = \frac{1}{k_{21} + k_{12}(c_A + c_B)} \; , \qquad \frac{1}{\tau} = k_{21} + k_{12}(c_A + c_B) \quad . \tag{150}$$

The following also applies to the above equilibrium:

$$K = \frac{c_{AB}}{c_A \cdot c_B} = \frac{k_{12}}{k_{21}} \quad .$$

A relaxation time of $\tau = 3 \times 10^{-6}\,s = 3\,\mu s$ results from Eq. (150) if $K = 10^5\,l\,mol^{-1}$, $c_A = c_B = 10^{-5}\,mol\,l^{-1}$ and k_{12}, the diffusion-controlled rate constant of the observed association equilibrium, equals $10^{10}\,l\,mol^{-1}\,s^{-1}$.

This shows that the heating period must be of the order of microseconds if we are to be able to follow this fast reaction. Therefore, the temperature jump must occur extremely rapidly. This is achieved by discharging a condenser in the actual sample cuvette [66]. Depending on the details of the arrangement (voltage, capacitance of the discharge condenser, separation between the electrodes, resistance of the solution and volume of the cuvette) temperature jumps of 4 to 10 K can be achieved in a few microseconds with this technique. In order to keep the resistance of a solution as low as possible an inert electrolyte, which must not influence the reaction to be measured, is added. $NaClO_4$, at a concentration corresponding to an ionic strength of 0.1 M, is frequently used for this purpose. The resistance is ca. 400 Ω in a typical cuvette with 6 mm diameter electrodes, a 10 mm gap and a 0.1 M inert salt concentration.

In this case, the heating time is approximately 2 µs if a capacitance of 0.01 µF ($\tau = RC/2$) is assumed. Reich and Sutter have obtained time constants shorter than 175 ns. They designed a specific temperature-jump cuvette in which particular care was taken to reduce the internal resistance [70]. In another version the electrical energy was fed in via a coaxial cable which enabled heating pulses as short as 80 ns to be achieved [71].

In addition to the conventional means of supplying energy as Joule heating by an electric current, other possibilities have been proposed. Microwave [72] and laser heating [73] are particularly applicable if the conductivity of the sample solution is too low, i.e. when relaxation processes in non-electrolytes (non-aqueous solutions) are being investigated.

The technique described by Holzwarth et al. [73a] utilizes an iodine laser which has an emission wavelength of 1.315 µm with a pulse energy of 1 to 20 J. The pulse period is either 2.4 µs or 3 ns. This laser is particularly well suited to heating water as a solvent.

For the detection of the time dependent concentration changes following the temperature jump we use the Bouguer-Lambert-Beer law in the form:

$$A = -2.303 \cdot \ln \frac{I}{I_0} = \varepsilon_i d \cdot c_i \; . \tag{151}$$

In the case of a differential concentration change as a result of a temperature jump this becomes:

$$\Delta A = -2.303 \cdot \Delta \ln I = \varepsilon_i d \cdot \Delta c_i \; .$$

From this we obtain for Δc_i:

$$\Delta c_i = \frac{\Delta A}{\varepsilon_i d} = -\frac{2.303}{\varepsilon_i d} \cdot \frac{\Delta I}{I} \; . \tag{152}$$

The change of light intensity ΔI is measured by means of a photomultiplier and followed on an oscilloscope or stored in a data acquisition system as a function of time from the beginning of the temperature jump. Since the signal is observed as a voltage V and because $I \sim V$, we obtain from Eq. (152)

$$\Delta c_i = \frac{2.303}{\varepsilon_i d} \cdot \frac{\Delta V}{V} \; . \tag{153}$$

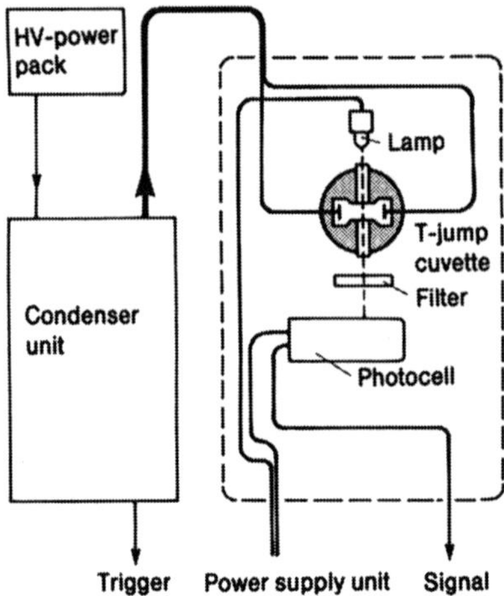

Fig. 70. Temperature-jump spectrometer TJ-1B, HI-TECH [74] for use in practical classes

Thus, the relative signal change is proportional to the change in concentration of the absorbing species, if $\Delta c_i/c_i$ is small.

Figure 70 shows schematically a temperature-jump apparatus which is suitable for teaching purposes [74]. The sample is placed between two electrodes in a cuvette. A condenser of 0.5 µF is charged with a voltage of 4 KV. The cuvette has a volume of 0.5 ml. The discharge supplies 4 J of energy which causes a temperature jump of ca. 4 K in 100 µs. Instruments developed for research work operate at higher voltages (ca. 20 KV) and smaller capacitances (0.04 to 0.01 µF) and provide temperature jumps of 6 to 10 K within a few microseconds. The time scale obtainable with these instruments covers $5\times10^{-6}-5\times10^{-1}$ s.

Proton-transfer reactions, metal-ligand reactions, redox reactions and enzyme-substrate reactions are examples of reactions which can be investigated by means of the temperature-jump technique. In the case of protolytic reactions which cannot be followed directly by spectrophotometric means or which lie in a spectral region unfavorable for measurement, the dissociation equilibrium under investigation can be coupled with an indicator equilibrium. In comparison with the previously described simple case, more complicated expressions for the concentration dependence of the individual components and for the relaxation times are obtained for this multistage-reaction system [66, 75].

The kinetics of proton transfer have been investigated intensively [76–79]. Of particular interest is the base catalyzed keto-enol interconversion of acetylacetone which was investigated in detail by Ahrens, Eigen, and others [78]. Measurements on acetylacetone solutions in the absence of a proton donor form a suitable system for introduction to this methodology and for testing equipment [74]. At this point, the descriptions in the original literature should be consulted [78, 80, 81].

A further example of the application of the temperature-jump technique is in the investigation of the kinetics of the reaction $Fe^{3+} + SCN^-$ which has already been mentioned (see the stopped-flow technique [82, 83]). This reaction has often been studied and the influence of a variable static pressure upon the observable rate constant has also been investigated [84, 85]. Jost has developed an appropriate cuvette for the application of the temperature-jump technique at variable pressures of up to 400 MPa [86]. Analogous investigations of the formation of a complex between Ni^{2+} and murexide in water [87], the hydrolysis of alizarine yellow GG with OH^- ions [88], and the self-association of dyes [89] have been carried out. Crooks has given a summary on the fast proton-transfer processes [90].

The temperature-jump method is suitable for following chemical reactions on the 10^{-6} s – 1s time-scale by spectroscopic means. However, in the vast majority of cases using the pressure-jump method, the reaction has been followed by the change of conductivity as a result of the sudden pressure change [67]. Knoche has reported a pressure-jump relaxation technique with spectroscopic detection [91]. Buschmann et al. have investigated the reversible hydration of carbonyl compounds by this technique [92].

In addition to these methods, the *field-jump technique* [93] should be mentioned. This utilizes the dissociation-field effect [94]. Here again, the electrical conductivity of an electrolyte solution is used as the detection method. De Maeyer has given a brief but clear description of all the possible techniques for investigating chemical relaxation processes [95].

Flash photolysis is another spectroscopic method for following fast reactions. In this method, short-lived substances are produced by means of an intense flash and the subsequent reactions followed directly spectrophotometrically. When using pulsed lasers, the pulse-width which initiates the reaction can be reduced to the nano- or pico-second region; and hence the kinetics which can be followed can be regarded as photokinetics.

Porter has given a description of flash photolysis [96].

7.6 Photoreactions

In addition to the 'dark' reactions discussed so far, photoreactions are of considerable importance to chemists. A photoreaction is initiated if molecule M first absorbs a light quantum of suitable energy $h\nu$:

$$M + h\nu \rightarrow M^* \, .$$

This results in an excited species M^*, which can subsequently start the photochemical reaction. In an ideal case, this reaction continues as long as light of appropriate energy is supplied. If the light is interrupted, the reaction stops, provided that no dark reactions occur as consecutive or reverse reactions. In this, the simplest case, the progress of the photochemical reaction can be followed in the dark phase by direct spectrophotometric means.

We can see immediately that the extent of a photoreaction will be quite generally a function of the light intensity and exposure time.

A photoreaction represents one possible deactivation process of an excited molecular state. Therefore, a photochemical reaction competes with those primary physical processes which we discussed in Sect. 2.2 by means of the energy-level diagram in Fig. 2.

Knowledge of the quantum yield is of critical importance in evaluating a photochemical reaction [97]. By definition we differentiate between different quantum yields:

a) an apparent total quantum yield,
b) a true total quantum yield,
c) an apparent differential quantum yield and
d) a true differential quantum yield.

Quantum yields a) to c) are generally dependent in a complex way on the period of irradiation. The *true differential* quantum yield d) is particularly important for the mathematical discussion of *simple photoreactions* [97].

Simple photoreactions are reactions in which only one substance initiates the photoreaction by light absorption and where all dark reactions occur so rapidly that we can apply Bodenstein's steady state hypothesis to the unstable intermediate products [98]. The other reaction partners can absorb light but they do not initiate a photoreaction. Photoreactions which do not meet both these criteria are called complex photoreactions.

The true differential quantum yield is defined as follows:

$$\psi_b^a = \pm \frac{\dot{b}}{I_a} . \tag{154}$$

$\dot{b} = d[b]/dt$ is the change of concentration with respect to time of component b.

I_a are the moles of light quanta per volume and time unit absorbed by substance a.

The notation, ψ_b^a, indicates that component a absorbs the light, i.e. it initiates the photoreaction and component b, whose change of concentration is then measured, is formed. In this case, the positive sign applies. However, if a is consumed and the reaction is measured via the decrease in a, the negative sign applies.

From Eq. (154) we obtain directly:

$$\dot{b} = \frac{d[b]}{dt} = \psi_b^a I_a . \tag{155}$$

Photoisomerization provides an example of a simple and uniform reaction whose quantum yield can be described as shown in Eq. (155):

$$a \xrightarrow{h\nu} b .$$

If we are dealing with more complex photoreactions it is appropriate to divide them into partial photoreactions which are themselves defined by time-independent partial quantum yields.

$$\psi_k^{a_i} = \frac{\dot{x}_k}{I_{a_i}} , \quad \frac{dx_k}{dt} = \dot{x}_k = \psi_k^{a_i} \cdot I_{a_i} . \tag{154a}$$

In this equation, x_k is the extent of reaction of the kth partial reaction and a_i the component which initiates the photoreaction by light absorption [5].

The + and − signs refer to substances k which either form (+) or disappear (−).

Although Eqs. (154), (155) and (154a) are simple, an easy integration is not possible since the quantity of light absorbed, I_a, depends in a complex way on the reaction site and irradiation time. Following Mauer [5, 97], and making certain assumptions, we can obtain from Eq. (155) the expression:

$$\dot{b} = 1000 \, \psi_b^a \varepsilon_a' c_a I_0 \, \frac{1 - e^{-\sum_1^n (\varepsilon_i' c_i) d}}{d \sum \varepsilon_i' \cdot c_i} . \tag{156}$$

If the absorbance $A' = \sum \varepsilon_i' c d = 2.303 \, A$ is substituted for $\sum (\varepsilon_i' c_i(t)) d$ we obtain:

$$\dot{b} = \frac{d[b(t)]}{dt} = 1000 \, \psi_b^a \varepsilon_a' c_a I_0 \frac{1 - e^{-A'(t)}}{A'(t)} . \tag{156a}$$

We define the expression $(1 - e^{-A'})/A' = F(A')$ as the *photokinetic factor* which must be considered in all photokinetic equations if the quantum yield is independent of concentration and intensity.

If the quantum yield is to be determined as shown in Eq. (156a) then the following conditions must be met:

1. The reaction mechanism of a photoreaction must be known;
2. the extinction coefficients must be measureable at the radiation wavelength λ_e (note $\varepsilon' = 2.303 \, \varepsilon$!);
3. it must be possible to follow the concentration change of the components under observation;
4. it must be possible to determine the light-source intensity at the surface, I_0, in Einstein/cm^2 s in order to determine the quantity of light absorbed, I_a, at wavelength λ_e.

As indicated in condition 4, when following a photoreaction, the intensity, I_0, of the light source used must be determined at wavelength λ_e. This can be done by either physical or chemical methods.

Although physical detectors such as a photoelement, photodiode, photocell or photomultiplier make possible a simple measurement of radiation intensities, an absolute measurement always entails great effort [100, 101]. For this reason, *chemical actinometers* for measuring the radiation intensity have proved to be valuable in photochemistry. Gauglitz has detailed the requirements which must be met by a *chemical* actinometer [100, 101].

The best-known chemical actinometer is the iron-oxalate actinometer of Parker and Hatchard [102, 103]. The experimental procedures for measurements with this actinometer have been described in detail in the literature [104–106].

In this actinometer, upon exposure to light, an Fe(II) salt is formed from an Fe(III) salt in acid solution with a quantum yield of $\psi_{\mathrm{II}}^{\mathrm{III}} \simeq 1.25$ which is almost independent of wavelength in the region 254–436 nm. After the photoreaction, the concentration of Fe(II) is determined photometrically at 510 nm via the 1,10-phenanthroline complex. This actinometer has the one disadvantage that the solutions must be mixed under red light and approximately one hour is needed to form the complex. Only one intensity value is established for each mixture which cannot be determined until sometime after the actinometry. This means that the extent of the reaction cannot be determined during the actinometry, which is the disadvantage of almost all classical actinometers.

More recent investigations into this actinometer have dealt with proposals for improving the practical procedures for carrying out the measurements [107, 108]. Bowman and Doemas [108] have pointed out the systematic errors of this actinometer.

For the above reasons, actinometers have been developed which allow the actinometric change to be followed directly using the same apparatus as for the measurement of the photometric reaction. They also allow, where possible, the acquisition of several intensity values for every actinometric measurement and the calculation of the result immediately after the actinometry. These requirements presuppose a uniform photoreaction which is as simple as possible.

A photoreaction which meets these requirements is the isomerization of *azobenzene* [109]. Mauser [36] and Gauglitz [110] have investigated this photoreaction in detail; thus, a suitable actinometer is now available for the region of 220–280 nm [111].

The reversible photooxidation of heterocoordianthrone (HCD) to its endoperoxide has been proposed for the region 300–370 nm [112]. In conjunction with investigations of azobenzene, Frank and Gauglitz have recommended 2,2′,4,4′-tetraisopropylazobenzene as an actinometric substance for the region 350–390 nm [113]. Other actinometric substances are:

Aberchrome 540: 450–550 nm [114];
HCD epoxide: 400–440 nm and 470–600 nm [115, 116];
Mesodiphenyl helianthrene: 470–620 nm [115, 117].

A quantum yield independent of wavelength has been established for the last two compounds. Furthermore, standard solutions are available commercially [118].

This summary shows that suitable actinometric substances are available covering the region 220–620 nm.

The investigation of the kinetics of a photochemical reaction is basically a three-step process:

I. Measurement of the reaction spectra in the dark phase as a function of the exposure time in order to detect isosbestic points as indicators of uniformity.
II. An evaluation of absorbances at several wavelengths as a function of the exposure time for the purpose of applying graphical matrix rank analysis for the determination of the number of linearly independent partial reactions.
III. Setting up and solving the concentration-time differential equations.

Steps I and II have already been detailed in Sect. 7.2. However, step III requires a short explanation since, as shown by the rate laws in Sect. 7.1, even a simple integration can encounter complications.

The solutions of the equations are so constituted that a graphical or numerical check is hardly possible without additional knowledge of the reaction parameters, e.g. the photostationary final concentrations or the rate constants. It has therefore proved worthwhile to apply the method of formal integration described in Sect. 7.3 to the solution of the rate laws for photochemical reactions [119].

Hezel [27] has shown that in practice a correction is required in the photokinetic factor expression. This correction takes account of the fact that the exciting light of wavelength λ_e leaving the cuvette is partially reflected at the exit window and subsequently suffers multiple reflections at the entrance and exit windows [120].

Thus the photokinetic factor becomes:

$$F'(A) = \frac{1-10^{-A}}{A} \cdot \frac{1-R}{1-R \cdot 10^{-A}} = F(A) \cdot r(A) \ , \tag{157}$$

where R is the degree of reflection at the quartz/air interface. We must also consider this in other relevant equations.

In cases in which the photoisomerization is reversible this must be taken into account in the analysis. This is the simplest case of a complex uniform photoreaction. A reverse reaction can occur as a dark reaction or as a photoreaction with a corresponding partial quantum yield. In the latter case

$$a \xrightarrow{h\nu} b \ ,$$

$$b \xrightarrow{h\nu} a$$

we can express the time-dependent concentration changes of “a” and “b” as:

$$\frac{d[a]}{dt} = -\psi_b^a I_a + \psi_a^b I_b ,$$

$$\frac{d[b]}{dt} = \psi_b^a I_a - \psi_a^b I_b .$$

The quantum yields ψ_b^a and ψ_a^b can be determined by the method of formal integration, as shown by the example of the photoisomerization of *trans*-azobenzene [27, 36, 121].

If the photoreaction of azobenzene is carried out in sulphuric acid then, after *trans-cis* isomerization, there follows as a dark reaction a further step; a photocyclization with subsequent dehydration to yield the aromatic ring system benzo(c)-cinnoline.

Azobenzene itself acts here as an oxidizing agent [122]. Mauser et al. have investigated this reaction in detail and they have been able to show that practically only the *cis*-azobenzene takes part in the dehydration [123]. The photolysis of azobenzene in a sulphuric-acid solution described in this work is a notable example of the kinetic analysis of a complex photoreaction.

Photocyclization reactions as a consequence of a photochemical *trans-cis* isomerization have frequently been investigated for stilbenes (→phenanthrene) [23, 123, 124] and its aza-derivatives such as 1-pyridyl-2-phenyl-ethene (→mono-azaphenanthrene) [125], dipyridyl-ethene (→diazaphenanthrene) [126], 1-pyridyl-2-(1-naphthyl)-ethene (→mono-azachrysene) [127], 1-pyridyl-2-(9-phenanthryl)ethene (→mono-aza-dibenzanthracene) [128], 1-pyridyl-2-4-quinolyl-ethene (→diazachrysene) [129], styryldiazine (→diazaphenanthrene, both Ns in one ring) [130].

Schulte-Frohlinde et al. have investigated the influence on photoisomerization of substitutents in the 4- and 4′-positions of stilbenes [131] in detail.

If we substitute the methine groups in the $-CH=CH-$ double bond, in turn, by nitrogen we arrive at azobenzene via the Schiff's base, benzalaniline. Photometric investigations of benzalanilines and their associated aza-analogs have been carried out [132]. Mauser [132] has made detailed studies of numerous other photoreactions such as sensitized photoisomerizations, photoadditions, photodimerizations, photoreductions, photoconsecutive reactions, photoparallel reactions as simple uniform, simple non-uniform, complex uniform and non-complex uniform photoreactions, together with their possible mechanisms and the basic kinetic equations derived from them [132].

In principle, practical measurements of the reaction spectra of a photoreaction can be made in the dark phase with any UV-VIS spectrophotometer. In using this *interval method* we interrupt the irradiation after a specific time and transfer the irradiated cuvette to a spectrophotometer. Then, we irradiate again for a specific time and so on. In addition to the reaction spectra $A = f(\lambda)$ the absorbance $A_{\lambda,e}$ of the solution at the radiation wavelength λ_e must be measured. With this mode of operation, the irradiation and measuring instruments are kept separate. However, use of this mode presupposes that in the dark phase no interfering dark reaction is superimposed on the photoreaction. It is basically better to use a combined measuring and irradiation apparatus. However, the sample chamber of the spectrophotometer needs to be re-designed appropriately. Mauser et al. [133] have suggested designs for an interrupted irradiation and a continous irradiation layout for a Zeiss PMQII spectrometer with a MM12 monochromator. They also point out the errors which can occur during such measurements.

For further examples of photochemistry, the relevant literature should be consulted [105, 134–137]. Murov's Handbook of Photochemistry contains extensive tables of numerous physical and specifically molecular-physical data which are important in photochemical reactions [138].

7.7 Spectrometers for Kinetic Measurements

As shown in Sects. 7.1–7.6, spectrophotometric methods are extremely well suited to the investigation of the kinetics of chemical reactions. Traditional equipment, on account of its optical design (monochromators), requires a finite time for measuring a spectrum. A recording speed of more than 100 nm min^{-1} should not be exceeded if we wish to avoid the influence of the mechanical inertia of the optical system on the form of the recorded spectrum. Thus, at least 3 min are required to record a spectral region of 300 nm. In the case of faster reactions it is thus not possible to record complete spectra during the course of a reaction since the composition of the reaction solution can change markedly within the measurement time of an individual spectrum. When using traditional equipment, we depend on measurements in the *time-drive mode* at a fixed wavelength. Information about the time-dependent course of the absorbance is obtained at several wavelengths using a fresh reaction mixture for each, preferably under identical conditions. However, it is always a problem ensuring the reproducibility of such conditions as the composition of a reacting solution of two components, the temperature, the period of mixing and the establishment of the start-time of the reaction. On account of unavoidable errors, the evaluation of complex reactions is very difficult. Furthermore, it is sometimes difficult to find all the wavelengths relevant to an evaluation because it can happen that the first partial reaction is completed by the time the first spectrum is measured.

Therefore, there has long been an interest in the development of fast recording UV instruments which can be used specifically for kinetic investigations.

7.7.1 Rapid Spectrometers

The traditional design of equipment, but with oscillating components, represented the first stage of development [139–147]. It is the movement of the grating in these instruments which determines the speed of measurement of a spectrum since a photomultiplier reacts extremely rapidly to signal changes. Thus, the functional principle is similar to routinely used traditional equipment, i.e. the optical layout has a light source, monochromator, sample and photomultiplier. The difference in these instruments lies in the design of the monochromator. Instead of a slow grating movement via an analog or stepping motor and a spindle, a mirror oscillates rapidly or a wheel with a set of angled mirrors rotates rapidly. As in the case of other equipment of this traditional pattern, the beam of light entering the *monochromator* is spectrally dispersed by a grating. Instead of rotating the grating, a movement which is slow because of the mass involved, a small mirror of low inertia directs the light, spectrally dispersed by the fixed grating, onto a detector via the exit slit. The time required for recording a spectrum can be controlled by the frequency with which the mirror is moved, and it lies in the region of a few milliseconds. The unavoidable cost of the mechanism is very high for use in routine equipment and computer control is also required. Therefore, instruments of this type are not widely available although they offer almost the same specifications with respect to optical resolution as the corresponding slow, traditional spectrometer.

7.7.2 FT-UV Spectrometers

In recent years, nondispersive instruments which operate according to the Fourier-transformation principle [148, 149] have gained importance in IR spectroscopy. The polychromatic light of a radiation source penetrates the sample and falls onto an interferometer instead of a monochromator. In practice a Michelson interferometer is generally used in which the rapid linear movement of a mirror produces the interference in the measuring beam. In order to calculate an appropriate spectrum in the frequency domain from an interferogram recorded by a suitable detector in the time domain by means of a computationally intensive Fourier transformation [150], a laser with highly monochromatic and coherent light is required as an auxilliary radiation source for calibrating the frequency axis.

Although FT-IR spectrometers have been widely available on the market for some time, commercial FT spectrometers for the UV region are a recent introduction.

The acquisition of optical spectra by interferometry with subsequent Fourier transformation is normally considered to offer three advantages over dispersive spectrometry. The Fellgett or multiplex advantage represents the efficiency of collecting data from the whole of the wavelength range during most of the duration of the experiment, rather than only from a limited wavelength band at any one time during the scan. The Jacquinot or throughput advantage represents the intensity gain from dispensing with slits, and the Connes or wavelength accuracy advantage results from the high wavelength stability of the reference laser. There may also be a benefit of data processing in the Fourier domain [151].

We speak of a multiplex-measuring method since the whole spectrum is recorded simultaneously by a receiver where the radiation of different wavelengths is encoded by a varying modulation [152]. If the detector is the main source of noise the multiplex method records only that noise level whilst the modulated radiation of all wavelengths is being summed as signal. This gives rise to the multiplex advantage whereby the signal/noise ratio is increased for every spectral element. In the UV-VIS region, noise is not detector limited [152, 153] and the multiplex advantage is lost [154a]: in some cases there may even be a multiplex disadvantage [154b]. Furthermore, the mechanical tolerances of the interferometer need to be comensurate with the wavelength, which makes fabrication expensive for FT-UV instruments. This, and the problem of a suitable calibration radiation source [155], has impeded the development of FT-UV spectrometers for a long time. However, equipment for building such instruments, even for the UV region (200 to 800 nm), has been described [156].

The trade off between sensitivity and resolution is such that, as in astronomical or in some atomic spectroscopy where extremely high resolution is required, Fourier transform spectroscopy offers substantial advantages. Chelsea Instruments Ltd., London have developed the Fourier-transform spectrometer FT 500 UV-VIS which can be used for measuring highly resolved emission and absorption spectra in the region 160 to 700 nm [157]. To date, no information is available concerning its use in kinetic studies.

A dispersive Fourier-transform spectrometer has no moving parts and is therefore attractive for situations where stability is paramount. The interferometer in this case disperses the radiation as an interferogram onto an area detector or array detector. Fourier transformation is then used to recover the spectrum. Two commercial spectrometers have been designed using such a system: the Groton HPLC detector [158] and the Hitachi U 6000 microspectrometer. The development of charge-transfer detectors [159] makes dispersive Fourier-transform instruments a more attractive proposition than hitherto.

7.7.3 Diode Array Spectrometers

Diode array spectrometers [160] incorporate modern data processing and detector [161] technology into the original measurement methodology of spectroscopy. Figure 6, Sect. 3.2, shows and describes the optical layout of such a spectrometer. In contrast to dispersive recording UV-VIS double-beam spectrometers, diode array spectrometers measure all absorbance values of an absorption spectrum in a selected wavelength region simultaneously, i.e. at exactly the same point in time. This has an important advantage for kinetic measurements because, during the measurement of a reaction spectrum (see Sect. 7.2, Fig. 62) with a conventional recording double-beam instrument each extinction value is measured at a different time $t+\Delta t$, i.e. the chronological zero or starting point of a reaction cannot be determined exactly.

In such cases we proceed by deducting the absorbance values of the 1st and also 0th (initial) reaction spectrum from the absorbance values of the 2nd, 3rd...nth reaction spectrum. There is also an assumption that the measurement is made at the same recording speed and the same time intervals. This correction is not necessary in the simultaneous measurement of a reaction spectrum with a diode-array spectrometer. Furthermore, the lack of moving parts in the monochromator ensures excellent wavelength stability [162]. For these reasons, such an instrument is ideally suited for kinetic studies [163 a]. Several manufacturers produce diode array spectrometers, not only for kinetic measurements but also for general use and particularly as chromatographic detector systems. Figure 71 shows a comprehensive arrangement for kinetic measurements in the form of a block diagram. The UV-VIS diode-array spectrometer, Lambda 3840 from Perkin-Elmer, has been linked to a computer, PE 7500, which controls the spectrometer and collects the data [163 b].

Under the control of the Perkin-Elmer PECUV program, the measurements supply data at 0.25 nm intervals. Thus, for an average measurement range of 250 to 300 nm and a data format of 4 bytes per measuring point, ca. 4–5 kbytes of data per spectrum are obtained. However, for a meaningful kinetic evaluation 70 to 100 spectra per reaction are required, depending on the complexity of the reaction under investigation. Thus, for ar-

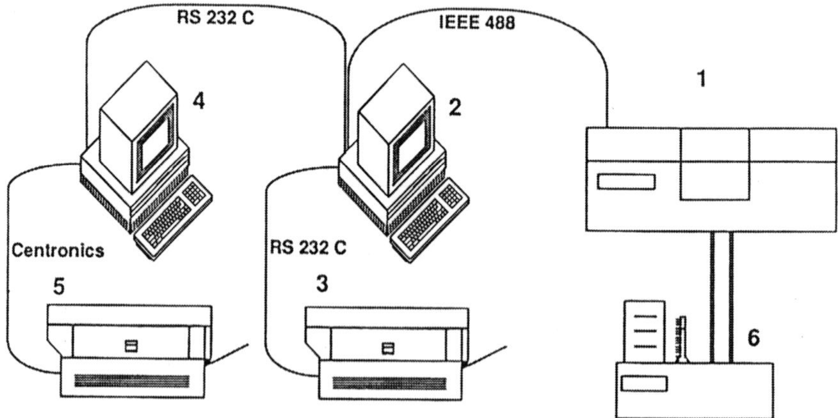

Fig. 71. Arrangement for kinetic measurements (block diagram). *1* Diode array-Spectrometer, Perkin-Elmer Lambda 3840; *2* Perkin-Elmer PE Computer 7500; *3* Epson FX 85 Printer; *4* Computer IBM PC XT; *5* Olympia Compact NP Printer; *6* Thermostat Haake FJ

chiving the measurements, it is advisable to reduce the data points per spectrum by one half, in order to store a complete reaction spectrum sequence on a 320 kbyte disk. Not all wavelengths are used for the evaluation. Wavelengths showing large absorbance maxima, or large absorbance changes during the course of the reaction are selected. These data reductions are made by the controlling computer, PE 7500.

In order to relieve the controlling computer of most of the evaluating tasks, selected reaction spectra or evaluation files with reduced data from each reaction are transferred from control computer, PE 7500, to the evaluation computer, IBM PC XT, via a serial interface. This has the great advantage that the next measurement can be made with the spectrometer whilst the last measurement is being evaluated. Furthermore, the results of the evaluation can be used in the selection of the measurement parameters for further experiments.

A comprehensive software package permits all relevant calculations and graphical displays necessary for a complete kinetic analysis to be run on the IBM PC XT [163 b]. The lower limit of the attainable time range is ca. 30 s. For longer periods (up to 10 h) the stability of the whole system (including the reaction solution) is the limiting factor.

For very fast reactions, i.e. reactions with half-life values below 2 s, the equipment described above cannot be used. For this another setup is required. The mixing of the sample components must be made using a flow method and the stopped-flow method is the most appropriate. Reactions which proceed even faster can only be investigated by means of the relaxation methods described in Sect. 7.5.2. In addition to time-drive measurement with a classical spectrometer which is always applicable, a diode array spectrometer can be used as the detector. Carl Zeiss, Oberkochen have developed a spectrometer unit which has no movable components and is optimized for very fast measurements [164]. Four different options are available and the spectral region to be investigated must be determined before purchasing the equipment since the enclosed system permits no subsequent modifications. If a particularly fast, IBM compatible, Turbo AT computer with a clock frequency of at least 10 MHz is used as control unit, 100 to 200 spectra per second are recordable. Very fast reactions in the ms region can be measured.

7.8 Determination of the Spectra of Intermediates

Information about the individual components of a reaction also forms part of a complete kinetic analysis. If the spectra of the individual components are known conclusions can often be drawn about their structure and statements made about the reaction mechanism. Several experimental and numerical methods can be used for determining the spectra of individual components.

A direct experimental determination utilizes all the options of modern analytical methods. If the reaction mixture is passed down an HPLC column the separation into individual components can be followed if suitable separating conditions are chosen [165, 166]. If a diode array spectrometer, see Sect. 7.7 [167–170], is used as detector, complete UV spectra can be measured during the short interval during which the eluate flows through the detector. In the case of complete separation, these spectra correspond to those of the individual components in the eluting solvent.

However, experimental determination by HPLC is limited by a number of secondary conditions. Thus, this method applies only to sufficiently slow reactions since measuring a chromatogram generally takes several minutes. Furthermore, the application is mainly confined to photoreactions in the dark phase: the conversion continues within the column in the case of dark reactions. Other interfering secondary reactions may be induced by the separation system. In addition, the solvent used in the reaction is often not suitable as the solvent for the separation column and can even destroy the stationary phase [171, 172]. For example, strongly alkaline solvents cannot be used since they attack stationary phases based on silica gel. Thus, the solvent may need to be modified prior to a chromatographic separation and this can affect the spectral behavior of the sample solution or even the course of the reaction.

A simple graphical evaluation based on the E-diagram can be carried out directly on the data. The extinction coefficients of the individual components at different wavelengths are determined by the points of intersection of the tangents to the individual partial straight lines in the E-diagram [173–176]. The inferior accuracy of the method and the difficulty of automation are disadvantages.

The methods and techniques dealt with in Sects. 7.2–7.4 permit determination of the order of the reaction and the rate constants. If a reaction has been followed spectrophotometrically, other information about the spectral of individual components is coded in the reaction spectra. It is easily possible to obtain this information numerically.

After having determined the order of the reaction and the rate constants, the concentration of each component at any time can be calculated. The following relation applies for the individual concentrations:

$$c_i = f(k_1 \ldots k_n, t) \ . \tag{158}$$

Together with the experimentally measured absorbance-time relationship, we generally obtain an overdetermined linear equation system, see Sect. 4.2:

$$A_\lambda = \left(\sum_{i=1}^{n} \varepsilon_{\lambda,i} \cdot c_i \right) \cdot d \tag{159}$$

or in matrix notation

$$\mathbf{A} = \varepsilon \mathrm{c} \mathrm{d} \tag{160}$$

$$\varepsilon \mathrm{c} = \mathbf{A}/\mathrm{d} \ . \tag{161}$$

The extinction coefficients may be determined by solving the system of equations using inverse multicomponent analysis. In the multicomponent analysis described in the literature, see Sect. 4.2, the extinction coefficients of individual components are generally known and the concentrations must be determined. In contrast, the task of inverse multicomponent analysis is the determination of the unknown extinction coefficients of individual components from the measured absorbance values and the known concentration-time relationship. The latter results from the closed solution of the basic differential equation system and the known parameters of the reaction, i.e. the rate constants and measurement times – via a generally overdetermined equation system for each individual wavelength. The overdetermination results from that fact that a reliable determination of the rate constants requires a considerably greater number of measurement points, i.e. times at which measurements are made, than the number of components. This applies particularly to complicated consecutive reactions. Whilst in a conventional multicomponent analysis the wavelength-dependent matrix of the extinction coefficients is multidimensional and the concentration vector is one-dimensional, in the case of an inverse multicomponent analysis there is in addition to the multidimensional matrix of the time-dependent individual concentrations a one-dimensional vector representing the unknown extinction coefficients. The values of these can be determined by means of the method of least squares:

$$\varepsilon \mathrm{c} \mathrm{c}^{\mathrm{T}} = (\mathbf{A}/\mathrm{d}) \mathrm{c}^{\mathrm{T}} \tag{162}$$

$$\text{with } \mathrm{D}' = (\mathbf{A}/\mathrm{d}) \mathrm{c}^{\mathrm{T}} \tag{163}$$

$$\text{and } \mathbf{c}' = \mathbf{c}\mathbf{c}^{\mathrm{T}} \tag{164}$$

we obtain the $\mathrm{n} \times \mathrm{n}$ equation system

$$\varepsilon \mathbf{c}' = \mathbf{D}' \ . \tag{165}$$

The spectral curve of each individual component can be obtained by setting up the equation systems for each individual wavelength and determining the appropriate extinction coefficients.

A modern digital computer executes this calculation simply and quickly [180]. This technique is superior to other methods since we do not have to interfere in the reaction system and, provided that the basic mechanism is correct, the exact absorbance can be obtained given sufficient numerical effort. For examples see [163b]. Modern spectrometers normally incorporate one of the many data processing packages now available, as discussed in Sects. 4.2 and 4.4.

References

1. Wedler G (1982) Lehrbuch der Physikalischen Chemie. Verlag Chemie, Weinheim Deerfield Beach, Florida, Basel
2. Barrow GM (1973) Physikalische Chemie, Gesamtausgabe. Bohmann, Heidelberg Wien; Vieweg, Braunschweig
3. Frost AA, Pearson RG (1964) Kinetik und Mechanismus homogener Reaktionen. Verlag Chemie, Weinheim
4. Schwetlick K, Dunken H, Pretzschner G, Scherzer K, Tieler H-J (1974) Chemische Kinetik, Fachstudium Chemie, Lehrbuch 6. Verlag Chemie, Weinheim
5. Mauser H (1974) Formale Kinetik. Bertelsmann Universitätsverlag, Düsseldorf
6. Mauser H, Polster J (1974) Z physik Chem NF 91:108
7. Fromherz H (1966) Physikalisch-chemisches Rechnen in Wissenschaft und Technik, 3. Aufl. Verlag Chemie, Weinheim, S 269 ff
8. a) Zachmann H-G (1972) Mathematik für Chemiker. Verlag Chemie, Weinheim, S 231 – 234
   b) Alexitz G, Fenyö St (1962) Mathematik für Chemiker. Akad Verlagsges, Geest u Portig, Leipzig, S 184 – 191
9. Ebisch R, Fanghämel E, Habicher W-D, Hahn R, Unverfehrt K (1980) Chemische Kinetik, Fachstudium der Chemie, Arbeitsbuch 6, 1. Aufl. Verlag Chemie, Weinheim
10. Derauleau DA, Dubler D (1981) Anal Biochem 114:411
11. Fahr E, Schmied M (1980) Fresenius Z Anal Chem 300:381
12. Ainsworth S (1961) J Phys Chem 65:1968; (1963) 67:1613
13. Hugus ZZ, El-Awady AA (1971) J Phys Chem 75:2945
14. Katakis D (1965) Anal Chem 37:876
15. Sternberg JC, Stillo HS, Schwandemann RH (1960) Anal Chem 32:84
16. Weber G (1961) Nature 190:27
17. a) Kowalsky HJ (1971) Lineare Algebra. de Gruyter, Berlin
    b) Zurmühl R (1958) Matrizen, 2. Aufl. Springer, Berlin Göttingen Heidelberg, § 7, S 83 ff
18. Magar ME (1972) Data Analysis in Biochemistry and Biophysics, Kap 9. Academic Press, New York
19. Bruhner JT, Shurwell HF (1973) J Phys Chem 77:256
20. Leggett DJ (1977) Anal Chem 9:276
21. Mauser H (1968) Z Naturforsch 23 b:1021, 1025
22. Mauser H, Polster J, Wenck H (1972) Chimia 26:361
23. Mauser H, Niemann HJ, Kretschmar R (1972) Z Naturforsch 27 b:1350
24. Mauser H, Starrock V, Niemann H-J (1972) Z Naturforsch 27 b:1353
25. Mauser H: loc cit [5], Kap IV, 2c, S 310 ff
26. Lachmann H, Mauser H, Schneider F, Wenk H (1971) Z Naturforsch 26 b:629
27. Hezel U (1969) Dissertation, Tübingen
28. Lachmann G, Lachmann H, Mauser H (1980) Z Phys Chem NF 120:19
29. Lachmann G, Lachmann H, Mauser H (1980) Z Phys Chem NF 120:9
30. Gauglitz G (1982) GIT Fachz Lab 26:205
31. Gauglitz G (1982) GIT Fachz Lab 26:597
32. Swinbourne ES (1975) Auswertung und Analyse kinetischer Messungen, Taschentext 37. Verlag Chemie, Weinheim
33. Swinbourne ES (1960) J Chem Soc 473:2371
34. Guggenheim EA (1926) Philos Mag (7) 2:538
35. Mauser H (1983) Z Naturforsch 38 a:359; (1964) 19 a:767
36. Mauser H, Hezel U (1971) Z Naturforsch 26 b:203
37. Lachmann H (1980) Fresenius Z Anal Chem 301:148
38. Lachmann G, Lachmann H (1978) Fresenius Z Anal Chem 290:118
39. Hezel U (1971) Zeiss Mitteil 5:316
40. Gauglitz G (1981) GIT Fachz Lab 25:537
41. Gauglitz G, Klink T, Lorch A (1981) In: Koch KH, Massmann H (eds) 13. Spektrometertagung. de Gruyter, Berlin New York, S 137 – 151

42. Asmus E (1959) Einführung in die Höhere Mathematik, 3. Aufl. de Gruyter, Berlin
43. Orlow YE, Piskareva RV (1974) Khim Prir Soedin 10:87; (1974) CA 80:119877h
44. Ursin B (1980) Diplomarbeit, Düsseldorf. Ursin B, Perkampus H-H, in Vorbereit.
45. Harbridge H, Roughton FJW (1923) Proc Roy Soc A104:376
46. Roughton FJW, Millikan GA (1936) ibid A115:258
47. Millikan GA (1936) ibid A155:277
48. Chance B (1940) J Franklin Inst 229:455, 737; (1949) J Biol Chem 179:1249; (1949) 180:865
49. SFA-11 Rapid Kinetics Accessory, HI-TECH, Scientific, Ltd. Vertretung Deutschland, AMKO GmbH, Gärtnerweg 49, D-2082 Tornesch
50. Chance B (1951) Rev Sci Instr 22:619
51. Gibson QH (1954) Disc Faraday Soc 17:137
52. Caldin EF (1964) Fast Reactions in Solution, Kap 3. Blackwell, Oxford
53. Hague DN (1971) Fast Reactions, Kap 2. Wiley, London
54. Roughton FJW, Chance B (1963) In: Fries SL, Lenis ES, Weinberger A (eds) Investigations of Rates and Mechanisms of Reactions, Kap 14. Interscience, New York
55. Chance B (1974) Investigations of Rates and Mechanisms of Reactions, 3rd edn. In: Hannes GG (ed) Techniques of Chemistry Services, Vol VI, Kap 2. Wiley, New York
56. Caldin EF, Crooks JE, Quenn A (1973) J Phys E (Sci Instruments) 6:930
57. Below JF, Counick RF, Coppel CP (1958) J Amer Chem Soc 80:2961
58. Lachmann H, Mauser H (1972) Hoppe-Seyler's Z physiol Chem 353:730
59. Lachmann H (1982) Habilitationsschrift, Univers Tübingen
60. Lübbers DW, Wodick R (1969) Appl Optics 8:1055
61. Santini RE, Milano MJ, Pardue HL (1973) Anal Chem 45:915A
62. Hollaway MR, Whik HA (1975) Biochem J 149:221
63. Papadakis N, Coolen RB, Dye JL (1975) Anal Chem 47:1644
64. Talmy Y (1975) Anal Chem 47:658A; Talmy Y (ed) (1979) Multichannel Image Detectors. Amer Chem Soc, Washington
65. HI-TECH Scientific Ltd s [49]
66. Czerlinsky G, Eigen M (1959) Z Elektrochem Ber Bunsenges Physik Chem 63:652; Czerlinsky G, Diebler H, Eigen M (1959) Z Physik Chem NF 19:246
67. Strehlow H, Becker M (1959) Z Elektrochem Ber Bunsenges Physik Chem 63:457
68. Eigen M, de Maeyer L (1963) In: Weißberger A (ed) Technique of Organic Chemistry, 2. Aufl, Vol VIII, 2. Interscience, New York
69. Ertl G, Gerischer H (1961) Z Elektrochem Ber Bunsenges Physik Chem 65:629
70. Reich RM, Sutter JR (1977) Anal Chem 49:1081
71. Hoffmann GW (1971) Rev Sci Instrum 42:1643
72. Caldin EF, Crooks JE (1967) J Sci Instrum 44:449; Buchwald HE, Ruppel H (1971) J Phys E: Sci Instrum 4:105
73. Caldin EF, Crooks JE, Robinson BF (1971) J Phys E: Sci Instrum 4:165; Turner DH, Flynn GW, Sutin N, Beitz JV (1972) J Am Chem Soc 94:1554; Aubard J, Meyer JJ, Dubois JE (1977) Chem Instrum 8:1
73. a) Holzwarth JF (1978) In: Gettins WJ, Wyn-Iones E (eds) Techniques and Applications of Fast Reactions in Solutions. Reidel Publishing Comp, Dordrecht, S 47–59; Frisch W, Schmidt A, Holzwarth JF, Volk R (1979) In: Gettins WJ, Wyn-Jones E (eds) Techniques and Applications of Fast Reactions in Solution. Reidel Publishing Comp, Dordrecht, S 61–70; Gruenwald B, Frisch W, Holzwarth JF (1981) Biochim Biophys Acta 641:311; Mareandalli B, Winzek C, Holzwarth JF (1984) Ber Bunsenges Phys Chem 88:368
74. Temperatur-Jump Spectrometer, Type TJ-1B; HI-TECH Scientific Salisbury, England, s. [49]
75. Bernasconi CF (1976) Relaxation Kinetics. Academic Press, New York
76. Eigen M (1963) Angew Chem 75:489
77. Ahrens ML, Maass G (1966) Angew Chem 80:848
78. Ahrens ML, Eigen M, Kruse W, Maass G (1970) Ber Bunsenges Physik Chem 74:380
79. Crooks JE (1975) In: Gold V, Caldin EF (eds) Proton-Transfer-Reactions. Chapman and Hall, London, S 153–177

80. Schwarzenbach G, Felder E (1944) Helv Chim Acta XXVII:1701
81. Eigen M, Ilgenfritz G, Kruse W (1965) Chem Ber 98:1623
82. Goodall DM, Havnson PW, Hardy JJ, Kirk CJ (1972) J Chem Educ 49:675
83. Cavasino FP, Eigen M (1964) Ricerca Scientifica 4:509
84. Jost A (1976) Ber Bunsenges Phys Chem 80:316
85. Doss R, van Eldick R, Kelm H (1982) ibid 86:925
86. Jost A (1974) ibid 78:300
87. Jost A (1975) ibid 79:850
88. Liphard KG, Jost A (1978) Ber Bunsenges Phys Chem 82:707
89. Ohling W (1984) ibid 88:109
90. Crooks JE (1975) In: Gold V, Caldin EF (eds): Proton Transfer Reaction. Chapman and Hall, London. Chap 6, 153–187
91. Knoche W, Wiese G (1976) Rev Sci Instrum 47:220
92. Buschmann H-H, Dutkiewicz E, Knoche W (1982) Ber Bunsenges Phys Chem 86:129
93. Eigen M, De Maeyer L (1955) Z Elektrochem Ber Bunsenges Phys Chem 59:986
94. Wien M, Chiele J (1931) Z Physik 32:545
95. De Maeyer L (1960) Z Elektrochem Ber Bunsenges Phys Chem 64:65
96. Porter G, West MA (1974) In: Hammes GG (ed) Investigations of Rates and Mechanisms of Reaction of Chemistry, 3rd edn, Vol VI. Wiley, New York
97. Mauser H (1967) Z Naturforsch 22b:367
98. Bodenstein M (1913) Z Phys Chem 85:329
99. Hellma GmbH, D-7840 Mühlheim/Baden
100. Gauglitz G (1978) Habilitationsschrift. Univers Tübingen
101. Gauglitz G (1983) Praktische Spektroskopie. Attempto, Tübingen, S 119ff
102. Parker CA (1953) Proc Roy Soc (London) A220:104
103. Hatchard CG, Parker CA (1956) ibid A235:518
104. Parker CA (1968) Photoluminescence of Solutions. Elsevier, London
105. Calverts JG, Pitts JN (1966) Photochemistry. Wiley, New York
106. Frank R, Gauglitz G Chemie u Anlagen, Verfahren, Juli 1978, S 19
107. Curien KC (1971) J Chem Soc B2081
108. Bowman WD, Demas JN (1976) J Phys Chem 80:2434
109. Stegemeyer H (1961) Dissertation. TU Hannover
110. Gauglitz G, Lüddeke E (1976) Fresenius Z Anal Chem 280:105
111. Gauglitz G, Hubig S (1981) J Photochem 15:255
112. Brauer HD, Drews W, Schmidt R: s Gauglitz, loc cit [101]
113. Frank R, Gauglitz G (1977) J Photochem 7:355
114. Heller HG, Langau JR (1981) J Chem Soc, Perkin I, 341
115. Gauglitz G, Hubig S (1982) Proc IX. JUPAC-Symp Photochem, Pau (France)
116. Brauer HD, Drews W, Schmidt R (1980) J Photochem 12:293
117. Brauer HD et al (1982) J Photochem 20:335
118. AMKO GmbH [49]
119. Niemann H-J, Mauser H (1972) Z Physik Chem NF 82:295
120. Mauser H, Francis DJ, Niemann H-J (1972) Z Physik Chem NF 82:318
121. Mauser H: loc cit [5], Kap IV, 4, S 353ff
122. Badger GM, Drewer RJ, Lewis GE (1966) Austral J Chem 19:643
123. Stegemeyer H (1962) Z Naturforsch 17b:153
124. Saltiel J, D'Agostino JT (1972) J Am Chem Soc 94:6445
Saltiel J, Chang DWL et al (1975) Pure Appl Chem 41:559
Saltiel J, D'Agostino JT et al (1973) Org Photochem 3:1
Saltiel J, Charlton JL (1980) cis-trans-Isomerisation of Olefine. In: de Mayo P (ed) Rearrangements in Ground- and Excited States. Academic Press, New York
125. Bartocci G, Mazzucato U, Masetti F (1980) J Phys Chem 84:847
Bortolus P, Cauzzo G, Mazzucato U, Galiazzo G (1969) Z Phys Chem NF 63:29
126. Whitten DG, Lee YC (1972) J Am Chem Soc 94:9142; Perkampus H-H, Kasseber G, Müller P (1967) Ber Bunsenges Phys Chem 71:40
127. Bartocci G, Mazzucato U, Bortolus P (1976/77) J Photochem 6:309

128. Aloisi GG, Mazzucato U, Spalletti A (1983) Z Phys Chem NF 133:107
129. Ursin B (1984) Dissertation. Univers Düsseldorf
130. Fehn H, Perkampus H-H (1978) Tetrahedron 34:1971
131. Bent DV, Schulte-Frohlinde D (1954) J Physic Chem 78:446, 451
Schulte-Frohlinde D, Bent DV (1974) Molec Photochem 6:315
Borrell P, Schulte-Frohlinde D (1975) Ber Bunsenges Physik Chem 79:662
Görner H, Schulte-Frohlinde D (1977) ibid 87:713
132. Mauser H: loc cit [5], S 137–166
133. Mauser H, Gauglitz G, Niemann H-J (1972) Z Physik Chem NF 82:309
134. Turro NJ (1965) Molecular Photochemistry. Benjamin, New York Amsterdam
135. Cundall RB, Gilbert A (1970) Photochemistry. Nelson, London
136. Becker HGO and Co-authors (1983) Einführung in die Photochemie, 2. Aufl. Thieme, Stuttgart, New York
137. Barltrop JA, Coyle JD (1978) Principles of Photochemistry. Wiley, Chichester New York Brisbane Toronto
138. Murov St L (1973) Handbook of Photochemistry. Dekker, New York
139. Lübbers DW, Wodick R (1969) Appl Optics 8:1055
140. Santini RE, Milano MJ, Pardue HL (1973) Anal Chem 45:915A
141. Hollaway MR, Whik HA (1975) Biochem J 149:221
142. Papadakis N, Coolen RB, Dye JL (1975) Anal Chem 47:1644
143. Dolin SA, Kruegle HA, Penzias GJ (1967) Appl Optics 6(2):267
144. Pimentel GC (1968) Appl Optics 7(11):2155
145. Babrow HJ, Tourin RH (1968) Appl Optics 7(11):2171
146. Denton MS et al (1976) Anal Chem 48(1):20
147. Ogan K, Essig A, Beach KW, Caplan SR (1977) Rev Sci Instr 48(2):142
148. Günzler H, Böck H (1983) IR-Spektroskopie. Verlag Chemie, Weinheim
149. Birch JR, Parker TJ (1979) In: Button KJ: Infrared and millimeter waves. Academic Press, New York
150. Cooley JN, Tukey JW (1965) Math Comp 19:297
151. Sharp MR (1984) Anal Chem 56:339A
152. Schrader B (1980) Infrarot- und Ramanspektroskopie in Ullmann's Enzyklopädie der Technischen Chemie. Verlag Chemie, Weinheim
153. Luc P, Gerstenkorn S (1978) Appl Optics 17:1327
154. a) Thorne AP (1991) Anal Chem 63:57A
b) (1990) Spectroscopy World 2(2):22
155. Connes P, Michel G (1975) Appl Optics 14:2067
156. Burton NJ, Mok CL, Parker TJ (1985) Optics Comm 45(6):367
157. Snook R, Grillo A (1986) American Laboratory 18:28
158. Bormann SA (1985) Anal Chem 57:276A
159. Bilhorne RB, Sweedler JW, Epperson PM, Denton MB (1987) Appl Spectr 41:1114
Vanderpant L, Taylor S (1991) Spectroscopy World 3(2):12
160. Burgess C, Knowles A (1981) Practical Absorption Spectrometry. Chapman and Hall, London
161. Jones DG (1985) Anal Chem 57:1057A, 1207A
162. Berlot PE, Locascio A (1991) Analyst 116:313
Burgess C (1987) In: Burgess C, Mielenz KD (eds) Advances in Standards and Methodology in Spectrophotometry. Elsevier, Amsterdam
163. a) Kaufmann R (1987) Dissertation. Universität Düsseldorf
b) Perkampus H-H, Kaufmann R (1991) Kinetische Analyse mit Hilfe der UV/VIS-Spektroskopie. Verlag Chemie, Weinheim
164. Mächler M, Schlemmer H (1989) Zeiss Inform 30, Heft 100:16
Wannowius KJ et al (1985) GIT Fachz Lab 29:1138
165. Gauglitz G, Klink T, Schmid W (1984) Fres Z Anal Chem 318:296
166. Gauglitz G, Klink T, Schmid W (1984) Fres Z Anal Chem 318:298
167. George SA, Maute A (1982) Chromatographia 15(7):419
168. Fell AF, Scott HP, Gill R, Moffat AC (1982) Chromatographia 16:69

169. Bormann SA (1983) Anal Chem 55(8):836
170. Fell AF, Scott HP, Gill R, Moffat AC (1983) J Chr 273:3
171. Engelhardt H (1977) Hochdruck-Flüssigkeits-Chromatographie. Springer, Berlin Heidelberg New York
172. Meyer V (1984) Praxis der Hochleistungs-Flüssigkeitschromatographie. Diesterweg-Salle-Sauerländer, Frankfurt
173. Mauser H, Starrock V, Niemann H-J (1972) Z Naturforsch 27b:1354
174. Mauser H, Gauglitz G (1973) Bunsen-Ber 106:1985
175. Ursin B (1984) Dissertation. Universität Düsseldorf
176. Polster J (1975) Z Phys Chem NF 97(3/4):9
177. Zachmann HG (1977) Mathematik für Chemiker. Verlag Chemie, Weinheim New York
178. Ebert K, Ederer H (1985) Computeranwendungen in der Chemie. Verlag Chemie, Weinheim
179. Herrmann D (1985) Numerische Mathematik. Vieweg, Wiesbaden
180. Miller AR (1982) Pascal-Programme Mathematik, Statistik, Informatik. Sybex, Düsseldorf

8 Evaluation of UV-VIS Spectral Bands

In Chapter 2, the basic theoretical correlations of electronic excitation spectra were briefly discussed and items of important molecular physical information which can be obtained from these spectra were pointed out. The relevant parameters are the absorption maxima corresponding to the term differences (= excitation energies) from the ground state; the intensities, given in ε_λ which correlate with the oscillator strength f; and the structure of the absorption bands traceable to overlapping of the vibrational excitation structure. Exact experimental data are required for correlation with theoretical calculations. However, UV-VIS absorption spectra measured in solution are often unstructured and have only weakly characterized shoulders which frequently make a clear assignment of excitation energies $E = hc \times \tilde{\nu}_{max}$ difficult or even impossible. A determination of the oscillator strength, f, from Eq. (5) is unreliable in these cases. For that reason, some important complementary assessments will be briefly described in this chapter.

8.1 Oscillator Strength and Transition Moment

The oscillator strength, f, the intensity of an electronic or vibronic absorption band, is defined [1, 2] as shown in Eq. (5) as:

$$f = \frac{2.303\,mc^2}{\pi e^2 N_L n} \int_{band} \varepsilon_{\tilde{\nu}}\, d\tilde{\nu} \tag{166}$$

where [see also Eq. (5)] e and m are the charge and mass of the electron, c is the velocity of light, N_L is the Loschmidt number, n is the refractive index and $\varepsilon_{\tilde{\nu}}$ is the molar decadic extinction coefficient at wavenumber $\tilde{\nu}$ in $l\,mol^{-1}\,cm^{-1}$.

The integral in Eq. (166) is to be extended over the appropriate electronic excitation bands and their vibrational structure.

Inserting the fundamental constants into Eq. (166) we obtain the relationship:

$$f = \frac{4.39 \cdot 10^{-9}}{n} \int_{band} \varepsilon_{\tilde{\nu}}\, d\tilde{\nu}\ . \tag{167}$$

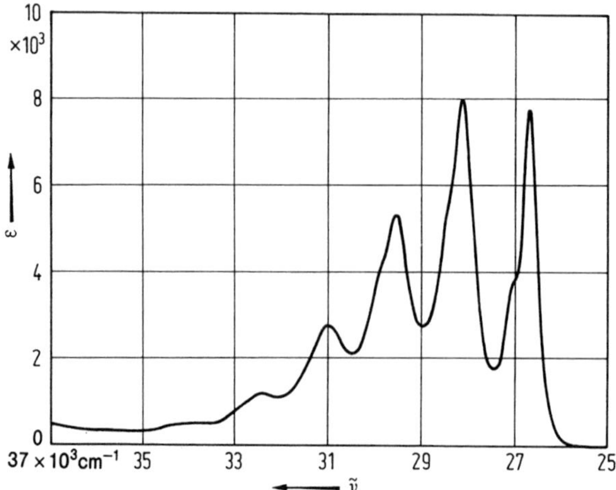

Fig. 72. ^1L_a band of anthracene; solvent: methanol; $c = 10^{-4}$ M; d = 1 cm; Perkin-Elmer 320

In practical applications, the refractive index is generally set equal to 1, i.e, ignored [3]. Establishing the oscillator strength requires an exact determination of the integral over $\varepsilon_{\tilde{\nu}} d\tilde{\nu}$.

Figure 72 shows the long wavelength absorption bands of anthracene, called the ^1L_a band in accordance with Platt's nomenclature [4], represented as $\varepsilon = f(\tilde{\nu})$. The integration of these bands in the range of $25000 \rightarrow 37000 \text{ cm}^{-1}$ was carried out in three ways:

1. The area enclosed by the absorption curve was determined by weighing the cut out area.
2. The area below the absorption curve was measured with a planimeter.
3. The area was determined by integration using the trapezium rule.

The integral is obtained in the units $|1 \text{ mol}^{-1} \text{ cm}^{-2}|$, since the numerical factor in (167) has the dimension 1^{-1} mol cm^2 and the oscillator strength is obtained as a dimensionless number.

Table 21 shows the results of the three evaluations:

The importance of the oscillator strength lies in the fact that it can be related theoretically to the transition dipole moment [1, 3, 6, 7].

For a transition from state l to state k we obtain for the oscillator strength:

$$f_{l,k} = \frac{8\pi^2 \cdot m \cdot c\tilde{\nu}_{l,k}}{3 \cdot h' \cdot e^2} G \, |\bar{M}_{l,k}|^2 \, . \tag{168}$$

$\tilde{\nu}_{l,k}$ is the wavenumber of the band center of gravity or the 0–0 transition and G is the statistical weight which may be set equal to "1" for electronic excitation. Inserting the numerical values of the fundamental constants we obtain by analogy with Eq. (166):

Table 21. $\int_{\text{bands}} \varepsilon_{\tilde{\nu}} d\tilde{\nu}$ and $f(^1L_a)$ for anthracene

Method	$\int \varepsilon_{\tilde{\nu}} d\tilde{\nu}$ [l mol^{-1} cm^{-2}]	$f(^1L_a)$ (159)	$f_{Lit.}$ [5]
Weighing	23.74×10^6	0.102	
Planimetry	23.11×10^6	0.0997	0.10
Numerical integration	22.44×10^6	0.096	

$$f_{l,k} = 4.70 \cdot 10^{29} \tilde{\nu}_{l,k} \cdot |\bar{M}_{l,k}|^2 \ . \qquad (169)$$

$\bar{M}_{lk}$ is the mean transition dipole moment in the units e.s.u. × cm.

With the oscillator strength value $f = 0.1$ determined above and the wavenumber of the $0-0$ transition for anthracene $\tilde{\nu}_{0,0} = 26600 \text{ cm}^{-1}$ we obtain from Eq. (169)

$$\text{Anthracene: } |\bar{M}_{l_{L_a}}| = 2.4 \text{ Debye} \ .$$

When the bands in an absorption spectrum can be clearly assigned to specific electronic transitions which show no overlap, the oscillator strength can be determined relatively easily from the data and a correlation with the theory is possible. In recent studies Klessinger [8] has dealt in detail with the question of this correlation and the quantum mechanical calculation of the oscillator strength.

Since the transition moment is a vector it can be represented by its components in the three directions in space:

$$|\bar{\vec{M}}_{l,k}| = \sqrt{\vec{M}_x^2 + \vec{M}_y^2 + \vec{M}_z^2} \ . \qquad (170)$$

If we designate the molecular plane with the coordinate axes x and y, then components $\vec{M}_x$ and $\vec{M}_y$ are orientated in this plane and $\vec{M}_z$ vertical to it. For planar molecules, the component $\vec{M}_z$ perpendicular to the $x-y$ plane can generally be neglected ($\vec{M}_z = 0$) for singlet transitions. In this case, the transition moment corresponds to the excitation of the electrons in the molecular plane. The direction of this transition moment then equals the resultant $\vec{M}_{l,k}$ of components $\vec{M}_x$ and $\vec{M}_y$ and we have:

$$\tan \alpha = \frac{\vec{M}_y}{\vec{M}_x} \ . \qquad (171)$$

In the event that one of the components M_x or M_y equals zero, we have an electronic transition polarized in one direction only and a characteristic anisotropy of the light absorption is observed.

This, for example, is the case for condensed aromatic hydrocarbons as shown by Coulson [9] and by Klevens and Platt [5] in their classic work.

For anthracene as an example, the assignment of the transition is as follows:

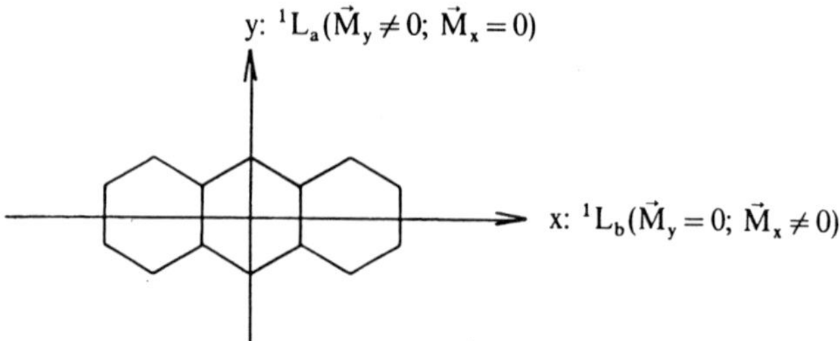

Accordingly, the ^1L_a transition (long wavelength bands) is orientated (polarized) along the short molecular axis (y) and the ^1L_b transition (forbidden transition) along the long axis (x). Craig and Hobbins [10] established experimentally that such an anisotropy of transition moments also applies to the intense absorption band at 40000 cm^{-1}.

The experimental detection of the anisotropy of electronic transitions, or the determination of the orientation of a transition moment relative to the molecular geometry, can also be carried out by means of absorption measurements using polarized light. However, in order to make these measurements we must know the orientation of the molecule relative to the plane of the polarization of the light or that it does not change during the measurements.

Molecular single crystals where all the molecules are arranged in parallel form ideal systems for these investigations. This particular case is found for many aromatic hydrocarbons [11], and this aspect of the spectra of aromatic single crystals was investigated at a very early date [12]. For this work a host crystal is often doped with a guest molecule and the spectroscopic properties of a "dissolved molecule", in the environment of the host molecules can then be investigated [13].

Dyck and McClure [14] used this method with a stilbene/dibenzyl mixed single crystal to show that the long wavelength band of *trans*-stilbene is polarized along the long axis.

Another readily applicable method for measuring the anisotropy of light absorption in conjunction with conventional UV-VIS measurements utilizes molecules dissolved in polymer films. If we stretch these solid films the long-chain polymer molecules are orientated in the direction of stretch whereby the dissolved molecules are likewise preferentially orientated in this direction. This orientation occurs particularly well with molecules which have a greater dimension in the longitudinal direction than in the transverse direction. Polyvinyl alcohol (PVA) and polyethylene are very suitable polymers for this technique [15–17].

Starting with a 10%–20% solution of PVA in water, to which the substance under investigation is added in aqueous or ethanolic solution, PVA films can be produced by pouring onto a glass plate. After evaporation of the excess water, films are obtained which are easy to handle and can be stretched after warming with a hair dryer.

For the actual absorption measurements we fit a polarizer (e.g. a Glan-Thomson prism or, for the visible region, a polarization filter) behind the exit slit of the monochromator. The stretched PVA film is fixed to the cuvette holder and the spectra are measured with the polarizer parallel and at right angles to the stretch direction, one after the other, against a pure unstretched reference film.

Scheibe et al. have tested this method with several cyanine dyes [15]. Perkampus et al. have determined the orientation of the transition moment for the long-wavelength bands of 1,2-dipyridyl ethene with stretched PVA films [16]. The absorption spectra of polyenes, cyanines and vitamin B_{12} were also investigated early on using stretched films [17].

Anthracene, its aza-analogs, acridine and phenazine [18], 9-methylanthracene and 1-hydroxaphenazine [19] have also been investigated in stretched PVA films. Hoshi et al. have described an extension of these investigations with stretched film to low temperatures. They modified a Shimadzu UV 360 spectrophotometer for this purpose [20].

In addition to PVA, polyethylene has proved to be useful. However, this is usually only available as a finished film. Therefore, another technique must be used for sample preparation. As described by Eckert and Kuhn [18], 0.1 mm thick polyethylene films are laid in a petri dish containing the almost saturated solution of the dye under investigation in chlorobenzene. After several days at room temperature an even distribution of the dye in the film is obtained. The dyed polyethylene film is washed with alcohol and stretched fivefold without warming.

Eggers and co-workers have carried out extensive investigations using this film technique [21]. They have also discussed the question of the extent to which the embedded molecules are orientated during the stretching procedure and how the orientation process can be explained in detail. Dekkers has also given a survey [22].

An older method should be described briefly. For very large molecules, i.e. if they have a large longitudinal extension, the anisotropy of light absorption can also be measured in a flowing solution. Scheibe introduced this method using a filiform polymer associate of a pseudo isocyanine dye in aqueous solution [23].

Another technique for investigating the anisotropy of light absorption has already been discussed in Sect. 5.5 in connection with luminescence excitation spectroscopy. However, in contrast to the methods described previously, these measurements require specially designed luminescence spectrophotometers [24]. Dörr has reviewed the methods of photoselection and the general theoretical and experimental principles of spectroscopy with polarized light [25]. A general description by Feofilov [26] should also be cited in this connection.

8.2 Band Analysis

8.2.1 Gaussian and Lorentzian Functions

It has already been mentioned in conjunction with the determination of the oscillator strength, that absorption bands are often very broad in the UV-VIS region and their more or less characteristic shoulders show that the observed bands consist of several overlapping sub-bands. In these cases, any statement about the position of the absorption maximum, λ_{max} or $\tilde{\nu}_{max}$, and the associated extinction coefficients ε_{max} is problematic. In such cases, the integrated extinction $\int \varepsilon_\nu d\nu$ is no longer a true measure of intensity since the degree of overlap cannot be determined. Vandenbelt and Henrich [27] have illustrated the effects of the summation of a pair of overlapping bands. The influence of two parameters, the relative intensity of the constituent bands and the separation $\Delta\lambda$ of their maxima, on the shape, observed band maxima and the total intensity within the band envelope were derived. An accurate statement about such an overlapping band system is only possible when the system is subjected to a band analysis to separate the partial and sub-bands.

In most cases, the bands to be separated can be approximated by Gaussian curves [28] and the extinction coefficient or absorbance can be written as follows:

$$A = A_{max} \cdot \exp\{-B(\tilde{\nu}_0 - \tilde{\nu})^2\} \tag{172}$$

($\tilde{\nu}_0$ is the wavenumber at the band maximum).

The parameter B is related to the full width at half the maximum intensity of the band by the equation:

$$B = \frac{4 \ln 2}{\Delta \tilde{\nu}_{1/2}^2} \; . \tag{173}$$

By forming the logarithm of Eq. (172) and taking the square root we obtain

$$\sqrt{\ln \frac{A_{max}}{A}} = \sqrt{B}\,\tilde{\nu} - \sqrt{B}\,\tilde{\nu}_0 \; . \tag{174}$$

This equation forms the basis for a linear regression for fitting the theoretical to the experimental curve. For the practical execution of this process it is necessary that:

There is a wavenumber (wavelength) interval in the spectrum which can be assigned to a single band only and which is not influenced by other bands.

Figure 73 shows the absorption spectrum of the dye morin in 0.1 M HCl. In this spectrum it can be seen that the long wavelength flank of the first absorption maximum can be assumed not to be influenced by the other

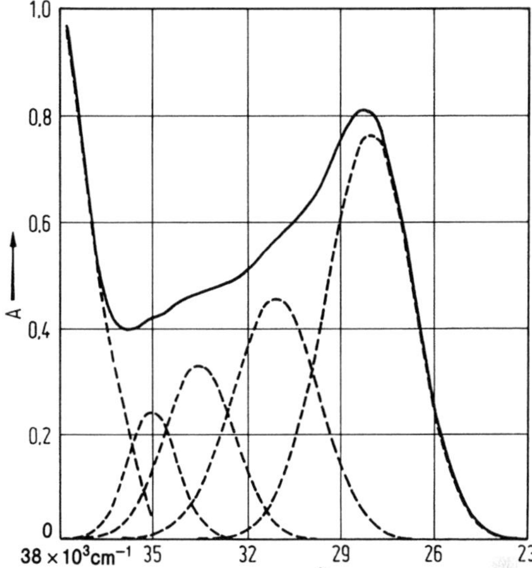

Fig. 73. Absorption spectrum of morin in 0.1 M HCl, 1st absorption band. Band analysis *dashed curve*; Perkin-Elmer 320; for the structural formula of morin, see Fig. 8

bands which can be recognized as shoulders towards shorter wavelengths, i.e. larger wavenumbers.

Furthermore, the steep drop of the absorption curve in the 27500–23000 cm^{-1} range indicates that the band should be approximated well by a Gaussian curve. For the absorption spectrum of morin in Fig. 73 the band analysis proceeds as follows:

We start with the maximum at the smallest wavenumber in the spectrum and look for the region on the first rising flank which is not influenced by other bands. Here we must vary three parameters:

1. the band maximum A_{max},
2. the starting wavenumber and
3. the final wavenumber of the selected interval.

The calculated absorbance values A are then fitted to the experimental curve by means of a linear regression corresponding to Eq. (174). The center of gravity of band $\tilde{\nu}_0(\lambda_0)$ is obtained from the intercept on the axis of the regression line. The value of the width at half the maximum intensity of the Gaussian curve is obtained from the slope of the regression line. The sum of the squared errors is determined for the curve. In the next step A_{max} and the wavenumber interval are varied. Every variation produces a new value of the error function:

$$F(A_{max}, \tilde{\nu}_1, \tilde{\nu}_n) = \sum_{i=1}^{n} (A_{i,exp} - A_i)^2 \colon \; [A(\tilde{\nu}_1) \ldots A(\tilde{\nu}_n)] \; .$$

A good fit is obtained if the error function has an absolute minimum.

After dealing with a wavenumber interval as described above we subtract the first extracted Gaussian curve from the measured spectrum and repeat the procedure on the rising flank of the second band which has now been exposed. Figure 73 shows that four subbands are found in the region of the illustrated absorption bands.

For asymmetric bands Eq. (172) can be extended by a cubic term [29]. The result in logarithmic form is then:

$$\ln\frac{A_{max}}{A} = b\cdot(\tilde{\nu}_0-\tilde{\nu})^2[1+a(\tilde{\nu}_0-\tilde{\nu})] \ . \tag{175}$$

Constants a and b can be determined from the curve. However, an error minimisation (regression) method should again be employed. $\tilde{\nu}_h$ is the wavenumber at which $A = A_{max}/2$; and $\tilde{\nu}_0 - \tilde{\nu}_h = \frac{1}{2}\Delta\tilde{\nu}_{1/2}$ so that the constant b can be expressed as:

$$b = B\frac{2}{1+a\Delta\tilde{\nu}_{1/2}} \ . \tag{176}$$

Again B is the constant defined by Eq. (173) and related to the width at half the maximum intensity of a pure Gaussian curve. Thus, in principle we can proceed in the same way as illustrated by the example.

In addition to the Gaussian function, the *Lorentzian function* is frequently used for band analysis. It is applied in the following form:

$$A' = \frac{a}{(\tilde{\nu}_0-\tilde{\nu})^2+b^2} \quad \left(A' = \ln\frac{I_0}{I}\right) \ . \tag{177}$$

We see that the following relationships exist:

$$\tilde{\nu} = \tilde{\nu}_0 \quad \text{and} \quad A' = A'_{max}$$

$$A'_{max} = \frac{a}{b^2}$$

and thus

$$A' = A'_{max}\cdot\frac{b^2}{(\tilde{\nu}_0-\tilde{\nu})^2+b^2} \ .$$

Since $A' = A'_{max}/2$ at $\tilde{\nu}_h$ and $\tilde{\nu}_0 - \tilde{\nu}_h = \frac{1}{2}\Delta\tilde{\nu}_{1/2}$ we have:

$$\Delta\tilde{\nu}_{1/2} = 2b \ .$$

Consequently, Eq. (177) can be written as follows:

$$A' = A'_{max}\frac{\Delta\tilde{\nu}_{1/2}^2}{4(\tilde{\nu}_0-\tilde{\nu})^2+\Delta\tilde{\nu}_{1/2}^2} \ . \tag{177a}$$

This relationship corresponds formally to the original Lorentz relationship for describing the shape of a band. The monographs of Kortüm [30] and Barker and Fox [30a] should be consulted for a more detailed discussion of the correlations.

8.2.2 Application of Derivative Spectra

Derivative spectroscopy, already discussed in Sect. 5.2, offers a further simple option for determining the number of overlying bands. It is also a good complement to, or check on, band analysis. Figure 74 shows the absorption spectrum of morin in 0.1 M HCl together with its 2nd order derivative spectrum. In contrast to Fig. 73, the spectrum was recorded in wavelength which was necessary because of instrumental constraints in the measurement of the derivative spectrum. Since the more intense bands below 270 nm also had to be measured, the maxima and minima of the long wavelength bands are relatively flat (see later for an explanation).

However, it can be seen quite clearly that at least three bands can be expected in the long wavelength region (420–270 nm). A fourth band, slightly above 280 nm, is indicated by an inflexion. If the ordinate in the 2nd order spectrum is expanded to $-0.1 < A < 0.1$ this becomes even clearer. In the short wavelength region, three very sharp bands can be recognized. Thus, we can localize 7 subbands in the absorption spectrum of morin by means of derivative spectroscopy.

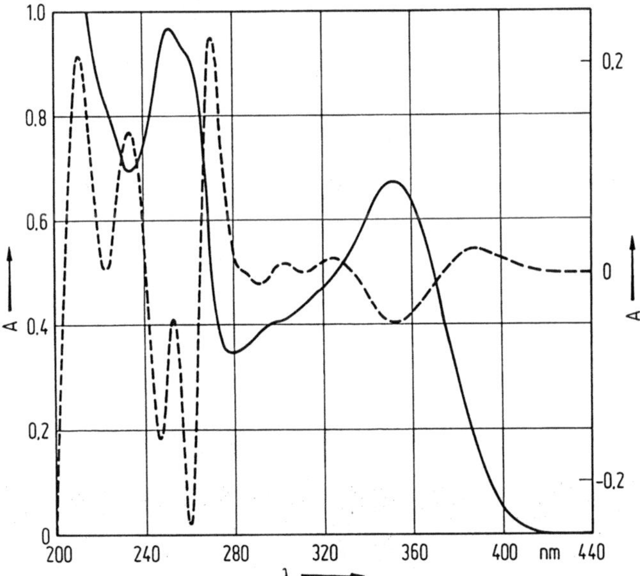

Fig. 74. Absorption spectrum of morin in 0.1 M HCl; *dashes*, 2nd order derivative spectrum, see text

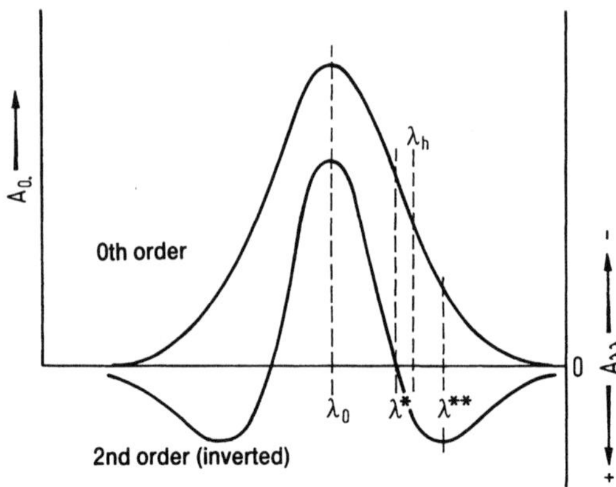

Fig. 75. Schematics of a Gaussian band in the 0th and 2nd order (inverted)

A comparison with the results of band analysis shows good agreement. However, a relatively large error function value compared with the other bands is found for the weak fourth band at ca. 35000 cm^{-1} (284 nm). This example provides evidence that it is very desirable to use these independent methods in combination.

Derivative spectroscopy also offers a simple possibility for establishing the relationship between the width at half maximum intensity and a 2nd order spectrum. Figure 75 shows the 0th order band as a typical Gaussian profile together with a 2nd order spectrum. The absorption maximum lies at λ_0 in the 0th order spectrum; λ_h denotes the wavelength which corresponds to half of the maximum absorbance or extinction coefficient. λ^* is the wavelength of the zero crossing of the 2nd derivative and λ^{**} corresponds to the wavelength where the spectrum of the 2nd derivative has its minimum value. If we formulate Eq. (172) in terms of the extinction coefficient ε and wavelength λ we can write:

$$\varepsilon = \varepsilon_{max} \exp\left\{-(\lambda_0-\lambda)^2 \cdot \frac{4\ln 2}{\Delta\lambda_{1/2}^2}\right\} \tag{178}$$

or with

$$B = \frac{4\ln 2}{\Delta\lambda_{1/2}^2} = \frac{\ln 2}{(\lambda_0-\lambda_h)^2}\ ; \quad \left(\lambda_0-\lambda_h = \frac{1}{2}\Delta\lambda_{1/2}\right) ,$$

$$\varepsilon = \varepsilon_{max} \exp\{-B\cdot(\lambda_0-\lambda)^2\} . \tag{178a}$$

If we differentiate Eq. (178a) twice with respect to λ, we obtain:

$$\frac{d^2\varepsilon}{d\lambda^2} = \varepsilon_{\lambda\lambda} = 2B\cdot\varepsilon[2B(\lambda_0-\lambda)^2-1] . \tag{179}$$

From this, the 3rd derivative is:

$$\frac{d^3\varepsilon}{d\lambda^3} = \varepsilon_{\lambda\lambda\lambda} = 4B^2\varepsilon(\lambda_0-\lambda)[2B(\lambda_0-\lambda)^2-3] . \tag{180}$$

Now, for the 2nd order spectrum, $\varepsilon_{\lambda\lambda} = 0$ at λ^*, and from Eq. (179) we obtain:

$$2B(\lambda_0 - \lambda^*)^2 = 1 \ , \quad (\lambda_0 - \lambda^*)^2 = \frac{(\lambda_0 - \lambda_h)^2}{2 \ln 2} \ ,$$

$$|\lambda_0 - \lambda^*| \sqrt{2 \ln 2} = |\lambda_0 - \lambda_h| = \tfrac{1}{2} \Delta\lambda_{1/2} \ . \tag{181}$$

$\varepsilon_{\lambda\lambda\lambda} = 0$ at the minimum of the 2nd order spectrum, λ^{**}, so that Eq. (180) gives:

$$2B(\lambda_0 - \lambda^{**})^2 = 3$$

or

$$|\lambda_0 - \lambda^{**}| \sqrt{\tfrac{2}{3} \ln 2} = |\lambda_0 - \lambda_h| = \tfrac{1}{2} \Delta\lambda_{1/2} \ . \tag{182}$$

Starting with Eq. (177a), we can write for a Lorentzian profile:

$$\varepsilon = \varepsilon_{max} \frac{1}{\left(\dfrac{\lambda_0 - \lambda}{\lambda_0 - \lambda_h}\right)^2 + 1} = \varepsilon_{max} [b(\lambda_0 - \lambda)^2 + 1] \tag{183}$$

with $b = (\lambda_0 - \lambda_h)^{-2}$.
The derivatives $\varepsilon_{\lambda\lambda}$ and $\varepsilon_{\lambda\lambda\lambda}$ are then given by:

$$\varepsilon_{\lambda\lambda} = \varepsilon_{max} \frac{2b[3b(\lambda_0 - \lambda)^2 - 1]}{[b(\lambda_0 - \lambda)^2 + 1]^3} \ , \tag{184}$$

$$\varepsilon_{\lambda\lambda\lambda} = \varepsilon_{max} \frac{24b^2(\lambda_0 - \lambda)[b(\lambda_0 - \lambda)^2 - 1]}{[b(\lambda_0 - \lambda)^2 + 1]^4} \ . \tag{185}$$

Since $\varepsilon_{\lambda\lambda} = 0$ at λ^* we obtain from Eq. (184):

$3b(\lambda_0 - \lambda^*)^2 = 1$ with $b = (\lambda_0 - \lambda_h)^{-2}$ whence it follows that:

$$|\lambda_0 - \lambda^*| \sqrt{3} = |\lambda_0 - \lambda_h| \ . \tag{186}$$

Analogously, since $\varepsilon_{\lambda\lambda\lambda} = 0$ at λ^{**}, Eq. (185) gives:

$$|\lambda_0 - \lambda^{**}| = |\lambda_0 - \lambda_h| \ . \tag{187}$$

If we summarize the results, we can see a simple relationship between λ_0, λ^*, λ^{**} and λ_h [31].

Gaussian curve:

$$|\lambda_0 - \lambda^*| = |\lambda_0 - \lambda_h| \cdot 0.85 \ , \tag{181}$$

$$|\lambda_0 - \lambda^{**}| = |\lambda_0 - \lambda_h| \cdot 1.47 \ . \tag{182}$$

Lorentzian curve:

$$|\lambda_0 - \lambda^*| = |\lambda_0 - \lambda_h| \cdot 0.58 \ , \tag{186}$$

$$|\lambda_0 - \lambda^{**}| = |\lambda_0 - \lambda_h| \tag{187}$$

or

$$\lambda^{**} = \lambda_h \; .$$

The distance between the band maximum λ_0 and the zero point of the 2nd order spectrum λ^* equals 0.85 times half the bandwidth at half maximum intensity for a Gaussian curve and 0.58 times for a Lorentzian curve. The minimum in the 2nd order spectrum equals $1.47\,|\lambda_0 - \lambda_h|$ for a Gaussian curve. In contrast, $\lambda^{**} = \lambda_h$ for a Lorentzian profile. On account of this fact, the bands in the 2nd order derivative spectra are clearer and easier to recognize.

It can also be seen that the relationships (181), (182) and (186), (187) are very useful in band analysis since, to a certain extent, they permit a differentiation between a Gaussian and Lorentzian profile.

Another relationship is obtained directly from Eq. (179). At λ_0, i.e. at the band maximum and at the minimum of the 2nd derivative we have:

$$\varepsilon_{\lambda\lambda}(\lambda_0) = -2\,\mathrm{B}\,\varepsilon_{\max} = -\frac{\varepsilon_{\max}}{(\lambda_0 - \lambda_h)^2} \cdot 2\ln 2 \; . \tag{188}$$

The minimum of 2nd derivative $\varepsilon_{\lambda\lambda}(\lambda_0)$ is inversely proportional to the square of half the bandwidth at half maximum intensity $\lambda_0 - \lambda_h$.

If two 0th order bands of similar intensity but of unequal width are compared then the relationship of their intensities in the 2nd order spectra derived from Eq. (180) is:

$$\frac{\varepsilon^{\max}_{\lambda\lambda,1}}{\varepsilon^{\max}_{\lambda\lambda,2}} = \frac{(\lambda_{0,2} - \lambda_{h,2})^2}{(\lambda_{0,1} - \lambda_{h,1})^2} \cdot \frac{\varepsilon_{\max,1}}{\varepsilon_{\max,2}} \; . \tag{189}$$

This is the basis of the fact that broad bands are suppressed in favor of narrower ones in a 2nd order spectrum. An example is shown by the spectrum of quinoline in *n*-heptane, Fig. 76. The 0th order spectrum shows the broad intense ^1L_a band next to the long wavelength, narrower, less intense ^1L_b band. In contrast, the inverted 2nd order spectrum shows only the well-structured ^1L_b transition whilst the ^1L_a band has almost completely disappeared. The same effect is responsible for the fact that in Fig. 74 the long wavelength bands in the 0th order spectrum of morin lead to weak signal intensities in the 2nd order spectrum whilst the short wavelength narrower bands are extremely well characterized.

The relationships derived for the 2nd order derivative spectra can also serve for the derivation of an expression for the oscillator strength. Equation (172) implies [multiplication of both sides by $(c \times d)^{-1}$]:

$$\varepsilon = \varepsilon_{\max} \exp\{-\mathrm{B}(\tilde{\nu}_0 - \tilde{\nu})^2\}$$

and thus

$$\int \varepsilon_{\tilde{\nu}}\, d\tilde{\nu} = \varepsilon_{\max} \int \exp\{-\mathrm{B}(\tilde{\nu}_0 - \tilde{\nu})^2\} d\tilde{\nu} \; .$$

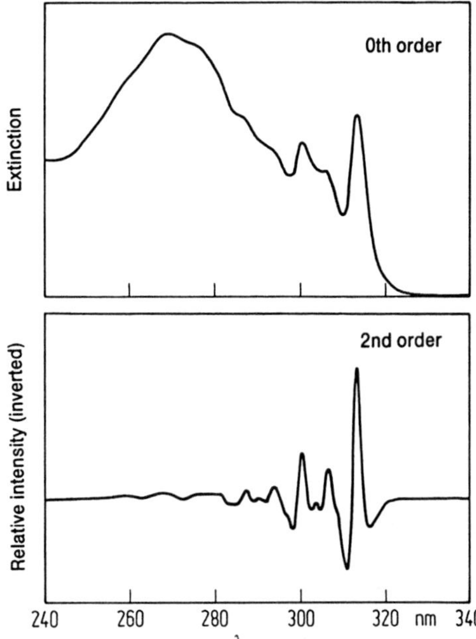

Fig. 76. 0th and 2nd order spectra of quinoline in *n*-heptane

If we change from wavenumber $\tilde{\nu}$ to wavelength λ the integral becomes

$$\int \varepsilon_{\tilde{\nu}} \tilde{\nu} \triangleq \frac{1}{\lambda_0^2} \int \varepsilon_\lambda \, d\lambda \ .$$

Introducing the Gaussian function of Eq. (178a) yields the integral

$$\frac{1}{\lambda_0^2} \int \varepsilon_{\max} e^{-B(\lambda_0-\lambda)^2} d\lambda \cong \varepsilon_{\max} \cdot \frac{1}{\lambda_0^2} \sqrt{\frac{\pi}{B}}$$

with

$$B = \frac{\ln 2}{(\lambda_0 - \lambda_h)^2}$$

it follows that

$$\frac{1}{\lambda_0^2} \int \varepsilon_\lambda \, d\lambda \sim \varepsilon_{\max} \frac{|\lambda_0 - \lambda_h|}{\lambda_0^2} \sqrt{\frac{\pi}{\ln 2}} \ .$$

From this, for the oscillator strength as shown in Eq. (167), there results:

$$f = 4.319 \cdot 10^{-9} \cdot \varepsilon_{\max} \frac{|\lambda_0 - \lambda_h|}{\lambda_0^2} \cdot \sqrt{\frac{\pi}{\ln 2}} \tag{190}$$

$$= k \cdot \varepsilon_{\max} \cdot \frac{|\lambda_0 - \lambda_h|}{\lambda_0^2} \cdot \sqrt{\frac{\pi}{\ln 2}} \, (k = 4.319 \cdot 10^{-9}) \ .$$

Substitution of ε_{max} from Eq. (188) provides the expression:

$$f = -\frac{k}{2\ln 2}\varepsilon_{\lambda\lambda}(\lambda_0)\frac{|\lambda_0-\lambda_h|^3}{\lambda_0^2}\cdot\sqrt{\frac{\pi}{\ln 2}}\,. \tag{191}$$

According to Eq. (181), we can substitute $|\lambda_0-\lambda^*|$ for $|\lambda_0-\lambda_h|$ and we obtain, since $|\lambda_0-\lambda_h| = \lambda_0-\lambda^*|\sqrt{2\ln 2}$:

$$f = -k\varepsilon_{\lambda\lambda}(\lambda_0)\cdot\frac{|\lambda_0-\lambda^*|^3}{\lambda_0^2}\cdot\sqrt{2\pi}\,. \tag{192}$$

In Eq. (191), and in the previous ones, the absolute values of the wavelength differences $\lambda_0-\lambda_h$, $\lambda_0-\lambda^*$, $\lambda_0-\lambda^{**}$ have always been used since λ_h, λ^* and λ^{**} can be greater or less than λ_0, but only the absolute values are important. As Fig. 75 shows, $\varepsilon_{\lambda\lambda}(\lambda_0)$ is always negative; see also Fig. 74.

We can conveniently use Eq. (182) to determine the ratio of the oscillator strengths from the 2nd order derivative spectra. This is of particular value when investigating factors which have an effect upon the intensity of the absorption spectrum of a compound; substituent effects and solvent interactions for example.

From Eq. (192), for two oscillator strengths, f_1 and f_2, the following applies:

$$\frac{f_1}{f_2} = \frac{\varepsilon^1_{\lambda\lambda}(\lambda_{0,1})}{\varepsilon^2_{\lambda\lambda}(\lambda_{0,2})}\cdot\frac{\dfrac{|\lambda_{0,1}-\lambda_1|^3}{\lambda^2_{0,1}}}{\dfrac{|\lambda_{0,2}-\lambda_2|^3}{\lambda^2_{0,2}}}\,. \tag{193}$$

The effect of the formation of a hydrogen bond on the long wavelength absorption bands ($^1A \rightarrow {}^1L_b$) of aza-aromatic substances provides an example of the application of Eq. (193) [31, 32].

8.3 Vibrational Structure

Numerous UV-VIS spectra measured in solution show a distinctive vibrational structure; the spectra of aromatic hydrocarbons (Fig. 17 anthracene, naphthacene, pentacene) or aza-aromatics (Fig. 18: 3,8- and 1–10 phenanthroline) for example.

Frequently the visibility of this structure is a function of the *solvent*. The *temperature* is also certainly important for the detection of structure since the vibrational structure becomes increasingly prominent at lower temperatures. For that reason, low-temperature cuvette holders have been developed which permit the measurement of UV-VIS absorption spectra in vitreous solidified solvents down to the temperature of liquid nitrogen [30, 33]. In some cases, these cuvette systems can be conveniently fitted to conventional UV-VIS spectrophotometers [33]. However, in-house construction is always to be recommended, since the design can be optimized to each individual spectrophotometer. A summary of suitable organic solvents which solidify to a glass, and the physical properties of these vitreous solvents and other matrices for low-temperature spectroscopy can be found in the work of Derkosch [34]. This summary is based on data from Scott and Allison [35] and from Meyer [36].

The vibrational structure of an absorption band generally shows a characteristic intensity gradation of the individual vibrational bands. However, two cases must be distinguished:

a) The first, long-wavelength vibrational band (0–0 transition) is the most intense, the intensity of bands lying at higher excitation energies decreases with increasing excitation energy.
b) The second or third vibrational band has the greatest intensity and the vibrational bands at lower or higher excitation energies have lower intensities.

An explanation of this intensity decrease is provided by the Franck-Condon principle which is a consequence of the Born-Oppenheimer approximation [1, 2]. The correlation between oscillator strength $f_{l,k}$ and transition dipole moment $|M_{l,k}|$ in Eq. (168) should accordingly be written more exactly as:

$$f_{l,k} = 4.701 \cdot 10^{29}\, \tilde{\nu}_{l,k} S^2_{10,kv'} |\vec{\bar{M}}_{l,k}|^2 \tag{194}$$

with the definition

$$\vec{M}_{l,k} = \vec{\bar{M}}_{l,k} \cdot S_{10,kv'} \ . \tag{194a}$$

In the spirit of the Born-Oppenheimer approximation we use an average value for the transition dipole moment [1, 2]. The term "$S_{10,kv'}$" is the overlap of the vibrational states participating in electronic excitation. Because of the Boltzmann distribution the overwhelming majority of molecules are in their lowest vibrational state, $v = 0$, at room temperature, and all transitions start from $v = 0$ of the electronic ground state "l". But they can reach vibrational states of $v' = 0, 1, 2 \ldots$ in the electronically excited state "k". A more exact theoretical treatment shows [1] that every vibrational sub-band $(0 \rightarrow v')$ of an electron transition $(l \rightarrow k)$ can be assigned an oscillator strength given by:

$$f_{l0 \rightarrow kv'} = \frac{f_{lk} \cdot \tilde{\nu}_{l0 \rightarrow kv'}}{\tilde{\nu}_{l,k}} |S_{l0 \rightarrow kv'}|^2 \tag{195}$$

where

$f_{l,0 \rightarrow k,v'}$ is the oscillator strength of the sub-band $l, 0 \rightarrow k, v'$ with any v' in the electronic excited state, k.

$f_{l,k}$ is the total oscillator strength of the transition $l \rightarrow k$ Eq. (186),
$\tilde{\nu}_{l,0 \rightarrow k,v'}$ is the wavenumber of the maximum of the respective vibrational band,
$\tilde{\nu}_{l \rightarrow k}$ is the wavenumber of the maximum of the 0–0 transition or center of gravity of the band,
$S_{l,0 \rightarrow k,v'}$ is the overlap integral of the vibrational function, $v = 0$, in the ground state, 1, with any vibrational function, v′, in the electronic excited state, k.

Thus, in the intensity gradation of the vibrational fine structure, the overlap integral is of crucial importance since it depends strongly on the geometry of the participating electronic states.

For the case in which the geometry does not change during the electronic excitation, i.e. the minima of the potential curves (or for triatomic or multiatomic molecules, the potential hypersurfaces) are exactly coincident, then the transition $l,0 \rightarrow k,0$ is the most intense since $S_{l,0 \rightarrow k,0}$ has reached its highest value. There is an optimum spatial overlapping of both participating vibrational functions for this *vertical* transition between the potential curves. This is caused by the fact that the most probable geometry of a vibrating molecule is always that where the internuclear distances are at their equilibrium values. According to the Born-Oppenheimer (Franck-Condon) approximation, the electronic transition occurs very rapidly. Thus, no account need be taken of a change of the positional coordinates of the nuclei during this time. Therefore, since we are considering the case of no change of geometry on excitation, state k with $v' = 0$ will also be at its most probable geometry, i.e. the participating vibrational functions are almost coincidental. According to this postulate, transitions to k with $v' \geqslant 1$ are only possible for molecules in the ground state with $v = 0$ which do not have the equilibrium geometry. However, there are always fewer of these. Thus, less molecules can make a vertical transition and consequently, the intensity is smaller.

In the second case, in which the geometry of the excited state differs from that of the ground state, the consequences can be seen immediately. Generally, it can be assumed that the equilibrium internuclear distances in the electronically excited state correspond to longer bond lengths. Thus, the maximum overlap integral $S_{l,0 \rightarrow k,v'}$ can only be achieved if the transition from $v = 0$ occurs to the higher vibrational state of $v' \geqslant 1$, but with the consequence that the transition $v = 0 \rightarrow v' = 0$ will show a lower intensity. Both cases of the Franck-Condon principle are shown in Fig. 77.

As shown in Fig. 77, it clearly follows that, in the evaluation of unresolved absorption bands, two characteristic band forms must, in principle, be distinguished. Spectrum a) corresponds the case of similar geometry in the ground and excited states.

The envelope of the vibrational sub-bands (dashed) shows a steep increase on the low wavenumber side and a flat spreading towards greater wavenumbers. On the other hand, in the second case spectrum b) produces

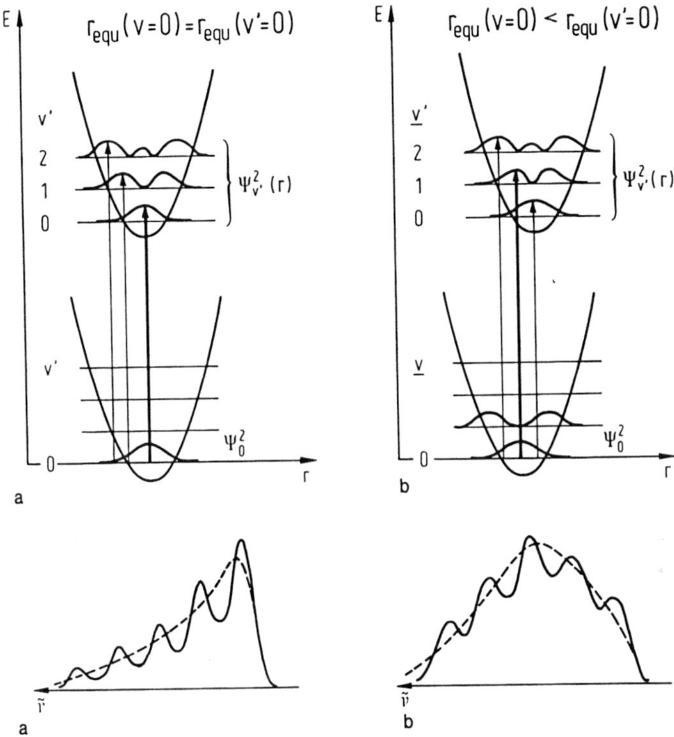

Fig. 77. a Illustration of the Franck-Condon principle, see text; **b** characteristic band shapes for unresolved absorption bands

a broad band (dashed) which is almost a Gaussian curve. From this it can be concluded that there is a different geometry in the ground and excited states. However, these conclusions can only be drawn if the UV-VIS spectra are plotted on a wavenumber scale since the overlap always corresponds to the overlap of unresolved excitation energies. A spectrum on a wavelength scale would therefore cause distortion of these band forms.

Equation (195) can be used directly in conjunction with the analysis of a vibrational structure. If $f_{1,k}$ is known for the whole band, then by determining $f_{1,0\rightarrow k,v'}$ for the individual, isolated vibrational sub-bands the overlap integral $S_{1,0\rightarrow k,v'}$ can be determined from the values $\tilde{\nu}_{1,k}$ and $\tilde{\nu}_{1,0\rightarrow k,v'}$ taken directly from the experimental spectra.

For an analysis of the vibrational structure in an absorption spectrum the question as to which normal vibrations couple with the electronic excitation must be posed. In some cases, this can be seen directly in the solution spectra. For example, the UV absorption spectrum of octatri-2,4,6-ine in *n*-heptane shows a very characteristic vibrational structure [37]. The individual maxima correspond to wavenumber differences of about 2100 – 2200 cm^{-1} which is comparable with the vibrational quantum of a $\nu_{C\equiv C}$ valence vibration.

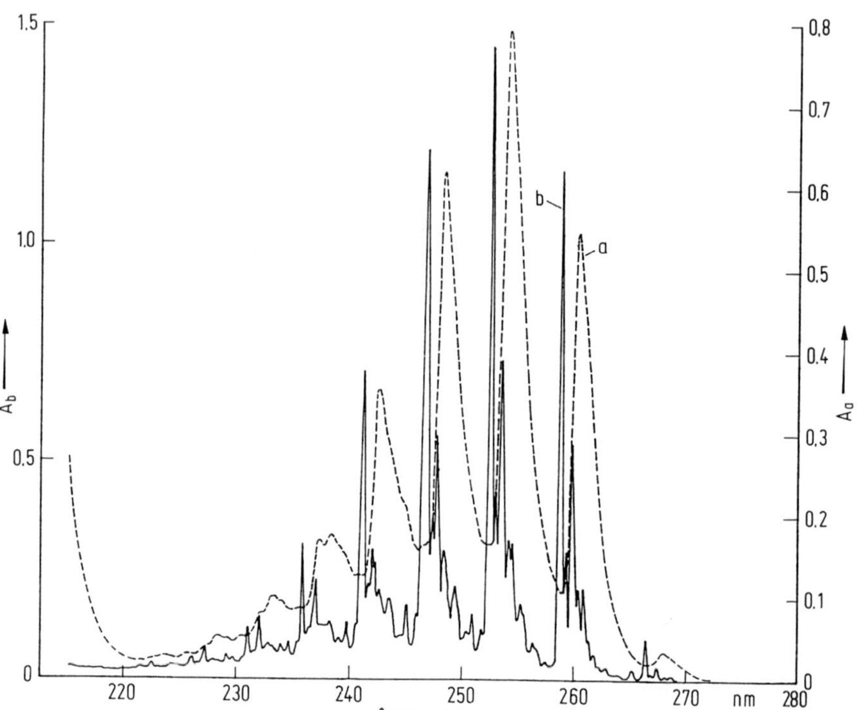

Fig. 78. Absorption spectrum of benzene (^1L_b band); *a* *n*-heptane solution, *b* vapor, Perkin-Elmer 320 instrument, $\Delta\lambda = 0.5$ nm

In the series of polyines $CH_3-|C\equiv C|_nCH_3$ with n = 2, 3, 4, 5 and 6 this progression is maintained [37]. In polyenes also, a wavenumber separation of about 1500–1600 cm^{-1} is observed [38, 39] which corresponds here to coupling of a C = C valence vibration with an electronic excitation.

In benzene, Fig. 78, a progression with a spacing of approximately 930 cm^{-1} is observed in the *n*-heptane solution spectrum. This corresponds to the totally symmetric breathing mode ν_2 of benzene which has been assigned in the Raman spectrum at 993 cm^{-1} [40]. When comparing these data, it must be noted that statements about the absorption spectrum relate to the vibrational quanta of the electronically excited state whilst the data from the IR or Raman spectra always apply to the electronic ground state.

Whilst absorption spectra generally show simple vibrational structure in solution at room temperature, the relationships become considerably more complicated if gas-phase spectra are measured. Figure 78b shows the gas-phase spectrum of benzene. In comparison with the solution spectrum, a multitude of very characteristic vibrational transitions are seen together with the dominant progression of ~930 cm^{-1}.

The vibrational structure of the gas-phase or rather, the vapor-phase spectrum of benzene shown in Fig. 78 was originally analyzed at an early date [41, 42]. Herzberg [40] and Murrell [2] have given accounts of this.

A comprehensive analysis of the vibronic coupling in UV-VIS absorption spectra requires a knowledge of the symmetry properties of the molecular electronic and vibronic states participating in the transition.

Here we must always differentiate between a *symmetry allowed* or *symmetry forbidden* electronic transition. Herzberg [43] has given a detailed description of the selection rules for the vibronic coupling in both cases. Further references may be found in those discussions of group theory in which spectroscopic applications are considered [44]. However, the analysis requires that not only is the vibrational spectrum of the relevant compound known but also the symmetry species of its normal vibrations. For this we generally require a normal coordinate analysis. Although normal coordinate analyses are frequently carried out, the number of molecules whose vibrational spectra have been exactly assigned is still small vis-a-vis the large number of molecules whose UV-VIS absorption spectra are known. The number of compounds which can be measured very exactly at high resolution as gas-phase or vapor-phase spectra is limited to those which can be transformed, undecomposed, into the gas or vapor state. The vast majority of compounds absorbing in the UV-VIS region have been measured in solution and such spectra generally show a simple vibrational structure. The assignment of the vibrational structure is then made purely empirically by comparison with known characteristic vibrational frequencies as in the examples above.

References

1. Birks JB (1970) Photophysics of Aromatic Molecules, Kap 3. Wiley, London New York Sydney Toronto, p 50ff
2. Murell J (1963) The Theory of the Electronic Spectra of Organic Molecules. Methuen, London
3. Mulliken RS (1939) J Chem Phys 7:14
4. Platt JR (1949) J Chem Phys 17:484
5. Klevens HB, Platt JR (1949) ibid 17:470
6. Förster Th (1951) Fluoreszenz organischer Verbindungen, IV § 12. Vandenhoeck & Ruprecht, Göttingen
7. Jaffeé HJ, Orchin M (1962) Theory and Application of Ultraviolet Spectroscopy, Kap 6. Wiley, London New York
8. Klessinger M (1982) Farbensymposium in Baden-Baden; und priv Mitteilung
9. Coulson CA (1948) Pr phys Soc (London) 60:257
10. Craig DP, Hobbins PC (1955) J Chem Soc 539
11. Robertson JM (1953) Organic Crystals and Molecules. Cornell University Press, Ithaca, New York
12. McClure DS (1959) Electronic Spectra of Molecules and Ions in Crystals. Academic Press, Solid State Reprints, New York London
13. Sidman J (1956) J Chem Phys 25:115, 122; ibid (1956) 24:757; McClure DS (1954) J Chem Phys 22:1668; (1956) 24:1

14. Dyck RH, McClure DS (1962) J Chem Phys 36:2326
15. Scheibe G, Kern J, Dörr F (1959) Z Elektrochem Ber Bunsenges Phys Chem 63:117
16. Perkampus H-H, Senger P, Kassebeer G (1963) Ber Bunsenges Phys Chem 67:703
17. Eckert R, Kuhn H (1960) ibid 64:356
18. Inoue H et al (1971) ibid 75:441
19. Hoshi T et al (1971) ibid 75:891
20. Hoshi T et al (1982) ibid 86:330
21. Eggers JH, Thulstrup EW, Have BP, Hansen LH, Swandström P (1965) Lecture at 8th European Congress on Molecular Spectroscopy, Copenhagen; Thulstrup EW, Eggers JH (1968) Chem Phys Letters 1:690; Thulstrup EW, Michl J, Eggers JH (1970) J Phys Chem 74:3868; Michl J, Thulstrup EW, Eggers JH (1970) ibid 74:3878; Ber Bunsenges Physik Chem 78:575; Thulstrup EW (1980) Aspects of the Linear and Magnetic Circular Dichroism of Planar Organic Molecules (Bertier G et al (eds) Lecture Notes in Chemistry, vol 14) Springer Verlag, New York
22. Dekkers JJ (1979) Academische Proefschrift. Freie Univers Amsterdam
23. Scheibe G (1938) Kolloid Z 82:2; Scheibe G (1966) in: Optische Anregung organischer Systeme. Z Internat Farbensymposium, Schloß Elmau, 1964. Verlag Chemie, Weinheim, S 109–142
24. Dörr F, Held M (1960) Angew Chem 72:287
25. Dörr F (1971) Polarized Light in Spectroscopy and Photochemistry in Creation and Detection of Excited State, vol I, A (Lamola AL, ed) Dekker, New York
26. Feofilov PP (1961) The Physical Basis of Polarized Emission, out of Russian. Consultants Bureau, New York
27. Vandenbelt JM, Henrich C (1953) Appl Spectroscopy 7:176
28. Kuhn N, Braun E (1930) Z Physik Chem B8:283
29. Siebert H, Linhard M (1957) Physik Chem NF 11:308
30. Kortüm G (1962) Kolorimetrie, Photometrie und Spektrometrie, 4 Aufl I, 7. Springer, Berlin Göttingen Heidelberg, S 47ff
30a. Barker BE, Fox MF (1980) Chem Soc Rev 9:143
31. Juffernbruch J (1982) Dissertation, Univers Düsseldorf
32. Juffernbruch J, Perkampus H-H (1983) Spectrochim Acta 39A:905
33. Lippert E, Luder W (1959) ibid 15:378
34. Derkosch J (1967) Absorptionsspektralanalyse im ultravioletten, sichtbaren und infraroten Spektralbereich. Akadem Verlagsges, Frankfurt/M
35. Cott DR, Allison JR (1962) J physic Chem 66:561
36. Meyer B (1971) Low Temperature Spectroscopy. American Elseviers, New York
37. Perkampus H-H, Bohlmann F (1966) In: Perkampus H-H, Sandemann U, Timmons CJ (Hrsg) DMS-UV-Atlas. Verlag Chemie, Weinheim. Butterworths, London, Vol I, A8/1 and A12/T1a
38. Ziegenbein W, in: DMS-UV-Atlas, A11/1 s loc cit [37]
39. Merz JH, Straub PA, Heilbronner E (1965) Chimia 19:302; Zeichmeister L (1960) Fortschr Chemie Org Naturstoffe 18:223; Weadon BC (1969) ibid 27:81; Vetter W, Englert G, Rigani N, Schwieter U (1969) ibid 27:189
40. Herzberg G (1966) Molecular Spectra and Molecular Structure, III. Electronic Spectra and Electronic Structure of Polyatomic Molecules. Van Nostrand, Princeton Toronto New York London
41. Sponer H, Nordheim G, Sklar AL, Teller E (1939) J chem Phys 7:207
42. Sponer H, Teller E (1948) Rev Med Phys 13:75
43. Herzberg G, loc cit [40], Kap II, 2, S 142ff; Herzberg G (1973) Einführ in die Molekülspektroskopie, Wissenschaftliche Forschungsberichte, Reihe IA, Bd 74. Steinkopff, Darmstadt, S 142ff; Garforth FM, Ingold CK (1948) J Chem Soc 417, 427, 433, 440; Garforth FM, Ingold CK, Poole HG (1948) J Chem Soc 491
44. Schonland D (1966) Molecular Symmetry. Benjamin, New York Amsterdam

Index of Illustrated Absorption Spectra

Subject Index

Printed by Printforce, the Netherlands